Understanding and Designing Computer Networks

Understanding and Designing Computer Networks

Graham A. King

Associate Professor in Computer Systems Engineering
Southampton Institute, UK

Edward Arnold
A member of the Hodder Headline Group
LONDON NEW YORK SYDNEY AUCKLAND

First published in Great Britain in 1995 by
Edward Arnold, a division of Hodder Headline PLC,
338 Euston Road, London NW1 3BH

British Library Cataloguing in Publication Data
A catalogue record for this book is available from the British Library

ISBN 0 340 61419 6

1 2 3 4 5   95 96 97 98 99

Typeset in Plantin by GreenGate Publishing Services, Tonbridge, Kent
Printed and bound in Great Britain by J.W. Arrowsmith Ltd., Bristol

Contents

Preface

The post industrial society makes enormous use of information. The infrastructure for this is computing and in particular computers that are able to exchange information. The links between machines must have form and protocol for operation. In defining form and protocol we create data networks.

The majority of current networks expect computers to have client-server relationships. One example of the server function is the file server. This describes one or more machines having a large backing store, usually magnetic disc or CD-ROM, upon which programs and data files reside. Other computers apply to these 'servers' to download the programs or data for execution or manipulation. Those who apply are the "clients". Access to other network facilities is often controlled through a server. Printing is sometimes provided through a print server which manages queues of documents.

Future paradigms of network operation will be different, with network parallelism created by systems that are able to use the network to distribute tasks or processes. They will achieve concurrent execution in a parallel computation environment.

Whether present or future models of networks apply, the key enabling technology is networking and it has never been more important to understand the principles involved.

This book aims to introduce the subject at a systems level and to demonstrate the application of networking principles. It achieves this integration of principle and practice by considering: a hypothetical company requirement; the possible design decisions that might apply; the resulting characteristics and later, practical matters of implementation and management. This text differs from many others in that it covers higher level processing issues such as data encryption and data compression whilst using the minimum of mathematics. It attempts to focus on intuitive underlying ideas because these topics are assuming greater importance as network security and network traffic related costs are becoming recognised as significant to network operations.

Network traffic is beginning to become more diverse and increasingly includes: speech; images; textual files and documents. The first phase in the wide ranging introduction of networks is over. The next phase will concentrate on internetworking and the efficient management of all aspects of networked integrated data systems.

GAK
1995

1

The requirement

1.1 A suitable case for study

Imagine the following scenario. A national car breakdown and rescue service recognises the need to introduce data networking into all levels of its operation. Sensibly, it is decided to proceed in a staged manner. The stages are agreed as:

1. Head office is to be networked so that desktop PCs can access data records, management information systems and electronic mail.
2. Branch offices are to be connected to head office so they may share features and be able to interact with data records.
3. Mobile units in the field are to be able to connect to the network in order to pass messages concerning their status and work.

Each of the stages will involve different technological solutions but they must be compatible. The outline specification for the first stage will be satisfied by the use of a local area network. The scope of local area networks has been loosely quantified as applying to distances up to 10 000 metres. This is sufficient for large buildings or sites.

1.2 The head office outline requirement

The network must carry relatively heavy traffic and estimates are that the signalling rate will need to be at least 4 megabits per second (Mb/s).

The head office is a tower block of four floors. Each floor is 200 metres, end to end. The floors are interconnected by stairwells and a lift shaft. Each floor is divided into offices and every point on every floor has to be potentially able to connect to the network. It is estimated that the maximum number of PCs per floor will be 100.

The system is to hold databases and the corporate data on these must be protected and secure from unauthorised access.

The project manager responsible for the creation of the network requires the network to be constructed in a modular fashion, allowing upgrading by parts. In particular the International Standards Organisation 7 layer model is to be used. Although most of these requirements are straightforward there is a great deal of freedom allowed in design and implementation. The requirements for stages two and three will be considered when appropriate but for now only the International Standards Organisation (ISO) model needs discussion.

1.3 The ISO system model

1.3.1 Rationale

Many network developers declare that they have complied with the ISO 7 layer model in creating their network. A set of jargon is associated with this model which endows a 'mystique' to the subject. This masks the reality, which is that the model is a tool for developers. It is a way of applying engineering discipline to the development task.

Developing a network is a complex job involving people with different of expertise. It is also a very large job. The standard engineering paradigm applied in such a case would be 'divide and conquer'. In other words, break the problem down into a series of smaller problems, allocate them to subsidiary teams of workers and be sure to define carefully the interfaces between the teams. In this way, when the various parts of the job are assembled, the overall task is achieved. If a standard form of breakdown were possible there would be less difficulty in adapting different developers' networks so that that they could talk to each other. To support this aim the International Standards Organisation (ISO) created an overall strategy which breaks the network design problem down into seven parts. Because these parts can be viewed as a hierarchy they are referred to by ISO as layers. Because adoption of the strategy is open to everyone and would facilitiate intercommunication between different networks the strategy is for 'open systems interconnection' (OSI). All this leads to the palindromic ISO OSI 7 layer model. To understand the model, a few example layers of the hierarchy will be considered in terms of the kinds of design issue addressed by each. No jargon will be used but a more comprehensive linking of jargon to functionality is dealt with in Appendix C.

1.3.2 Layer 1 – The lowest layer

The lowest layer in the hierarchy is concerned with signalling and the details of the data highway used by the network. The mechanical details of the cable or optical fibre used as the highway media will be defined by the sub-team working in this area. This will include: the plugs and sockets; the signalling levels, i.e. +12 V = 0, −12V = 1 or RS423 or something else; the cable (coaxial, twisted pair or what?). The whole purpose of the layer is to specify how the data is to be carried from place to place and it is only concerned with signalling 0s and 1s. There is no concern about the meaning of those 0s and 1s or whether any errors in transmission occur. It would be appropriate to employ electronics or communication engineers to implement layer 1.

1.3.3 Layer 2 – An error free link

Moving up the hierarchy it becomes necessary to ascribe some meaning to the 0s and 1s. This must be done in such a way that errors in transmission caused by electrical noise or other interference are able to be corrected. The usual strategy, and one employed by the ISO model, is to group 0s and 1s into blocks. These blocks of data will have 'fields' or parts reserved for network operational purposes. A network will have a number of users connected to a data highway via a 'station' or 'node'. Stations or nodes are interfaces, each of which is allocated a unique node number or 'address' on the network. In this way a message can be sent from node 3 to node 17, or from node 32 to node 8 and so on. The first field in a data block might carry the number of the destination node. The second might carry the number of the source. The third field might be used to number the block itself, so that it can be differentiated from others. The true message information might then follow in the 'data field'. After all of these, the final field might be a special code number, devised in such a way that it is possible to use the code to detect corruption of any data in the block. Details of how this could be achieved are covered in Chapter 4. Figure 1.1 shows the block format described. Data blocks are usually called 'data frames' and layer 2 is responsible for the data frame operation.

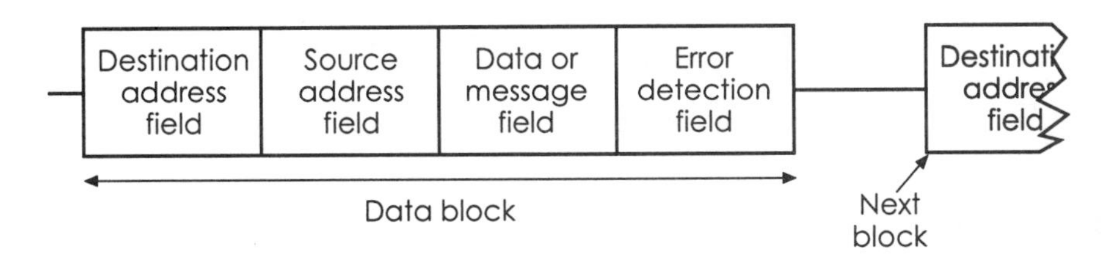

Figure 1.1 Data block format

To understand the overall operation consider one rule for data frame transfers. The rule states that each data frame must be acknowledged by the sending of an acknowledgement frame back from the destination to the source node. Other rules are possible but the rules for conversation are called 'the data transfer protocol' or simply the 'protocol'. Figure 1.2 demonstrates the working of the protocol when all is progressing well. Figure 1.3 illustrates the working when a data frame is corrupted.

The protocol action and the creation, sending and receiving of data frames is carried out by the nodes over the interconnecting data highway using the signalling defined by layer 1. Users simply supply messages to be inserted into data frames and simply receive messages extracted from data frames. The mechanism is invisible to the user, who would not know that some data was corrupted in transit and repeated, perhaps a number of times.

This means that the data frame operation, layer 2, is providing a service – an error free link.

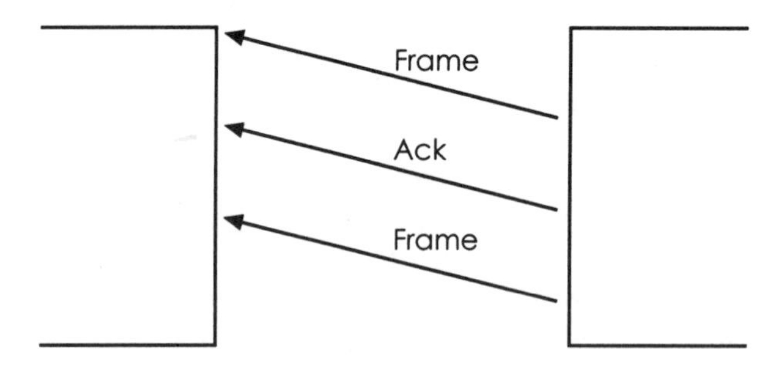

Figure 1.2 Protocol - data good

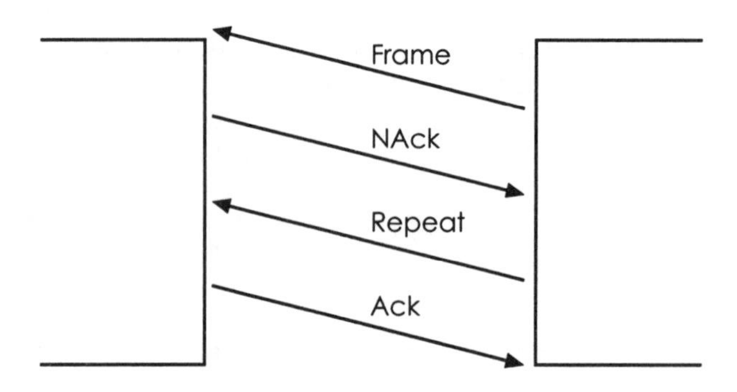

Figure 1.3 Protocol - data corrupted

1.3.4 Layer 3 – Data packets

In the same way that data frames are structured groups of bits,the message inside the message field may also be structured. Why? The answer begins with the realisation that one packet resides within one data frame as shown in Figure 1.4. Imagine that a data frame has been correctly received and that the node that has received it is expected to relaunch the message onto a different segment of the network. If the fields surrounding the message field have been stripped away at layer 2, how does the node know where to relaunch the message to? Bearing in mind that a relaunch implies the message being

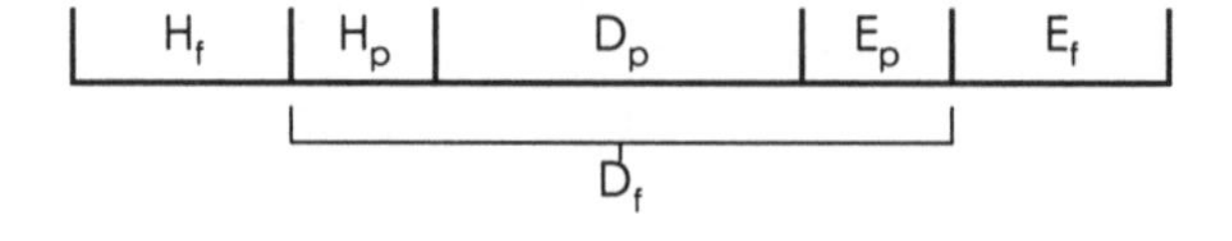

H_f = frame header (addresses etc.)
H_p = packet header
D_f = frame data field
D_p = packet data field
E_p = packet error correction
E_f = frame error correction

Figure 1.4 Frame and packet relationships

popped into a new data frame, what should be the destination address? This is a routing or re-routing problem and it is solved by the message having its own destination field. In fact it may have a whole set of fields just like a data frame. The re-routing node can now read the destination address field in the packet and create a new data frame appropriately.

1.3.5 Layer 4 – How the dialogue is achieved

One of the functions carried out by this subdivision of the network system is that of controlling the nature of the end-to-end dialogue. There are two basic options but only one would be chosen for a particular network. These options are best described by analogy.

When a telephone call is made the telephone system assumes a sequence of states. The first state involves setting up the call by accepting numbers and generating the ringing, engaged or unobtained tones. Once the call is connected the conversational state is attained and finally, one or other of the communicating parties clears down the circuit. 'Circuit' is the key word because the states correspond to establishing, using and then clearing the circuit. In data networks there is never a permanent circuit connection because data must be exchanged as a series of data packets/frames over a common highway. The actual data transfers are invisible to the user and so it is possible to gain the impression that a circuit does exist. A 'virtual circuit' is the term used to describe such a pseudo telephone call. Layer 4 software controls all phases of the call.

An alternative communication strategy is analogous to the sending of letters between two people. In this case, messages are self contained and are launched into the delivery system independently. Each must carry a full delivery address. This differs from the virtual circuit because in that case addresses are only necessary during the call set-up phase. The term used to describe the letter analogous system is 'datagrams'. Once again, layer 4 software would be responsible for operating such a link.

1.3.6 Layer 5 – Communication session control

If a virtual circuit had been established between two stations on a network and then, for unknown reasons the link was broken, how would communication be restored? In a telephone call the caller would redial and then carry on the conversation where it had been cut off. Layer 5 software provides the data link with the ability to carry on. It does this by marking transmissions with 'synchronising points'. There are usually major and minor points to create a structure as shown in Figure 1.5. This is best understood by imagining an archiving session, in which a sender is downloading the contents of a hard disc to a remote receiver. Minor synchronising points might be every file. Major points might be every directory. By marking these points it is easy for both ends to recover from interruption and carry on from the last recorded event or synchronising point. If this were not done for the example given it would be necessary to start all over again if there were a communication failure.

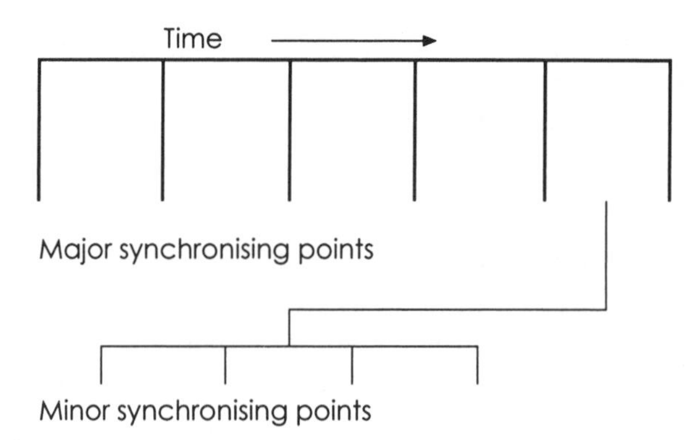

Figure 1.5 Synchronising points for session control

1.3.7 Layer 6 – Data processing and representation

Previous layers were functionally linked to the working of the network. Layer 6 is slightly different. It is one of the layers which acts to transform the nature of the data as it passes between the data source, say a computer, and the network. It could be that an application program running on the source operates on data in the form of numbers in binary coded decimal form, whereas the network only signals ASCII characters. The job of layer 6 is to convert both ways. There are other transformation processes that may be performed at this level, notably data compression/decompression, or data encryption/decryption, where security or confidentiality are a worry.

1.3.8 Layer 7 - Application software

Application software acts as an interface allowing applications processes, otherwise known as users' programs, to connect to the network.

1.3.9 The status of the model

The ISO 7 layer model is not original in the way that it suggests problems should be tackled. 'Divide and conquer' is a long established engineering strategy. Before ISO 7 was proposed, network developers had applied the general principle. They had generally produced slightly different layer splits and their models are often used in networks originating particularly from the USA. European products are more compliant with the ISO standard. Probably the best known alternative standards are those created by the American Institute of Electronic and Electrical Engineers (IEEE). The IEEE standard 802 committee created a number of model variants which do not correspond exactly to ISO 7 but do have equivalent functionality covering the lower layers. These IEEE standards will be mentioned again when the different possible local area layout shapes are considered in Chapter 3. There is a natural split in the implementation of the layers between layers 3 and 4. Up to layer 3 the likelihood is that hardware would provide the functionality. From layer 4 upward the implementation is all software. This gives an ideal 'cut' since ISO or other standards could interface to IEEE options at that point. A number of networks have been designed in this way.

Questions

1 A company decides to commission a local area network. As part of the requirements specification it is stated that the ISO 7 layer model is to be used in the development. Identify the purpose behind the use of the model and comment on the skills that design personnel require in order to implement the functionality of all the layers.

2 Explain the advantages of layered models from the point of view of the developer. Include in the discussion why it is important to allow independent development for the implementation of each layer.

3 In describing the ISO 7 layer model it is often said that there is peer to peer correspondence. What does this mean?

4 Discuss the concept of a protocol and generate a comprehensive list of the parameters and aspects that a protocol should define.

5 In layer 1 there is a possibility that RS standards could be specified. Explain the scope of this and other related EIA standards.

2

Data transmission

The signalling of computer words may be achieved by sending all bits of the word simultaneously over n lines. This is feasible for very short ranges and in cases where a computer is connected to an adjacent peripheral. A connection between a PC and a printer might be a typical example. Network specifications call for much longer range and usually define transmission media such as coaxial cable or twisted pair wires. In any case it would be uneconomic to provide n conductors over local area network distances of up to 10 km. The alternative is to signal the words one bit at a time. This requires a signal wire and a signal ground. A '1' might be signalled as a negative voltage level with respect to ground, while a '0' might be signalled as a positive level. This signalling is a function of time since each logic state must be sustained on the wire for a period. Figure 2.1 illustrates the concept. Signalling one bit at a time is serial communication. Receiving serial signals is a process that requires sampling of the line during the time for which a bit state is being sustained by the transmitter. Ideally the sampling instant is in the middle of the bit period and it is the means by which the exact moment is defined that classifies the signalling as either asynchronous or synchronous.

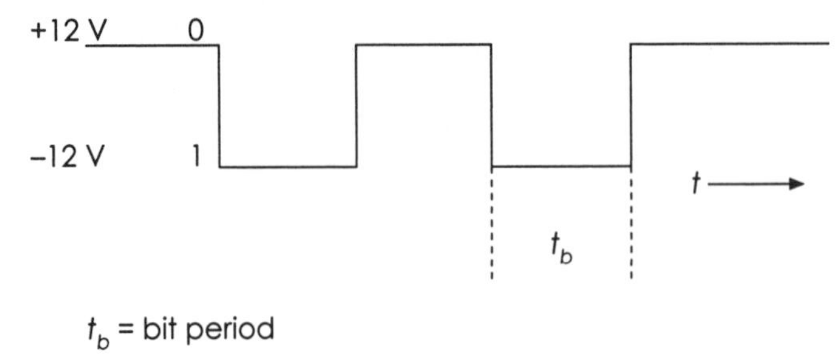

Figure 2.1 Serial signalling

2.1 Asynchronous Signalling

With asynchronous signalling, each byte is signalled as a separate entity. The eight information bits are framed by a start bit at the beginning and one or two stop bit periods at the end. When nothing is being signalled the line

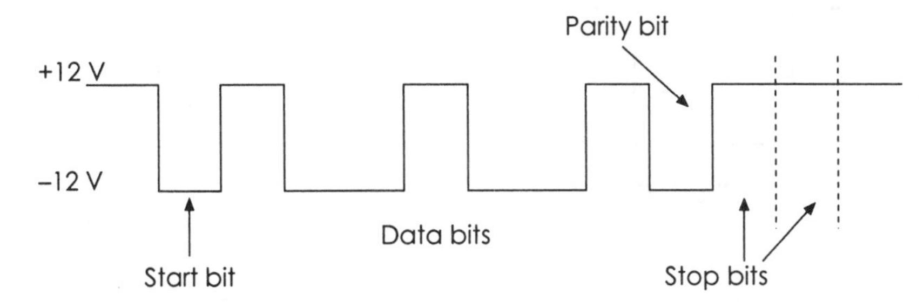

Figure 2.2 Start and stop bits

assumes an idle state. The idle state is usually the mark or '1' condition. The start bit represents a change from the idle state, whereas the stop bits are the same as the idle state. This strategy means that the start bit can be used to indicate that the signalling system is active and that the receiver must sample subsequent bits. The stop bits may be considered as an enforced idle state, allowing a new start bit to be unambiguously created. The effect is illustrated by Figure 2.2, but it should be noted that the information bits are signalled least significant bit first. Figure 2.3 shows why this is so. The transmitter clocks the byte out of a shift register least significant bit first in order that the receiver retains the correct sense of the byte after having clocked it in via the most significant bit of its own shift register.

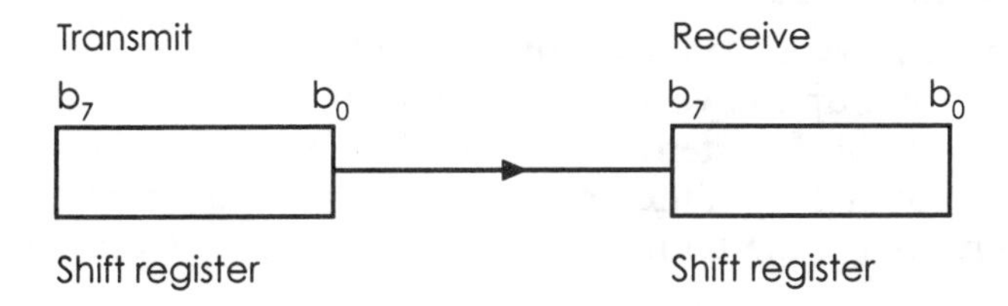

Figure 2.3 Rationale for signalling l.s.b. first

2.1.1 Sample timing

In the asynchronous system, timing is achieved through the use of timer-counter registers. Each end of the link bases its operations on a system clock, typically running at 16 or 64 times the bit signalling rate. With an X16 system, the clock is divided down by a /16 circuit to yield the bit rate for the transmit data register.

The receiver possesses a clock oscillator running at nominally the same frequency. There is no synchronism between the two ends, only a notional similarity. The receiver system also has a /16 divider but this divider is inhibited during idle signalling periods. On the arrival of the leading edge of a start bit, the receiver divider is clocked. This divider has available an output at count 8 and another at count 16. After count 8 is reached the signal line is sampled. If at '0', then the receiver continues and samples again every further 16 counts. The effect of this is shown in Figure 2.4.

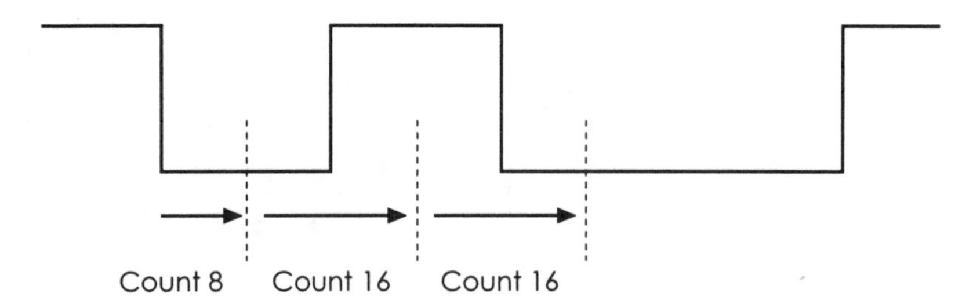

Figure 2.4 Sample timing - Asynchronous systems

Because a start bit leading edge can arrive at any time, irrespective of the current state of the receiver clock, there can be an error between the centre of each bit received and the count of 8 or 16. The maximum magnitude of the error is dependent on the divider ratio. In this case it will be 1/16 of a bit period. Since the clock oscillators at the two ends are independent there will be slight differences in their exact frequencies and this would result in a slowly increasing error in the sampling instant at the receiver. If allowed to continue the error would eventually become sufficiently large for the sampling instant to coincide with either a change of state or even the wrong bit. The problem does not arise because the asynchronous signalling system enforces an idle state with the stop bits after just one byte. The next start bit creates a new notional synchronism.

2.1.2 Asynchronous system overheads

Consider a typical asynchronous entity. It would comprise of a start bit, up to eight data bits, and up to two stop bits. Three bits in eleven are purely there to make the signalling system work. Another way of looking at this is to recognise that about 27% of the signalling bandwidth is wasted. What is more this figure is fixed. No matter how much data is signalled, the overhead is the same.

2.2 Synchronous Signalling

Synchronous transmission requires the receiver to be supplied with a clocking signal that is used to define the sampling instant. This clock is supplied by the transmitter and there are alternative ways by which it is provided. One option is for a separate line to carry the clock. It is inappropriate for a network to have a two channel main data highway and consequently most systems opt for a data signal which inherently carries clock information. The data clock is extracted from the serial data stream and used to sample the data stream itself. To understand this idea it is necessary to study binary modulation.

2.2.1 Binary modulation

The term binary modulation refers to a process of altering the form of a binary signal in order to guarantee the presence of particular characteristics. A complex wave can be analysed by Fourier's technique. If this is done it is found that the wave comprises a number of sinusoids which include a

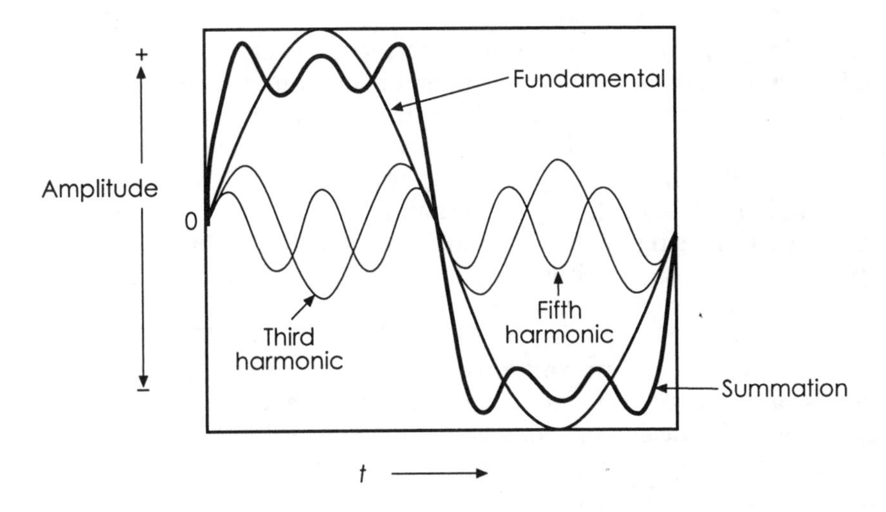

Figure 2.5 Harmonic synthesis

fundamental frequency and harmonics of the fundamental. For a rectangular wave train signalling 1010101010..., the equal mark-to-space ratio results in a fundamental sinusoid whose period is two bit periods. Only odd harmonics are present, e.g. three, five, seven times the fundamental. Figure 2.5 shows how these frequency components synthesise the wave train.

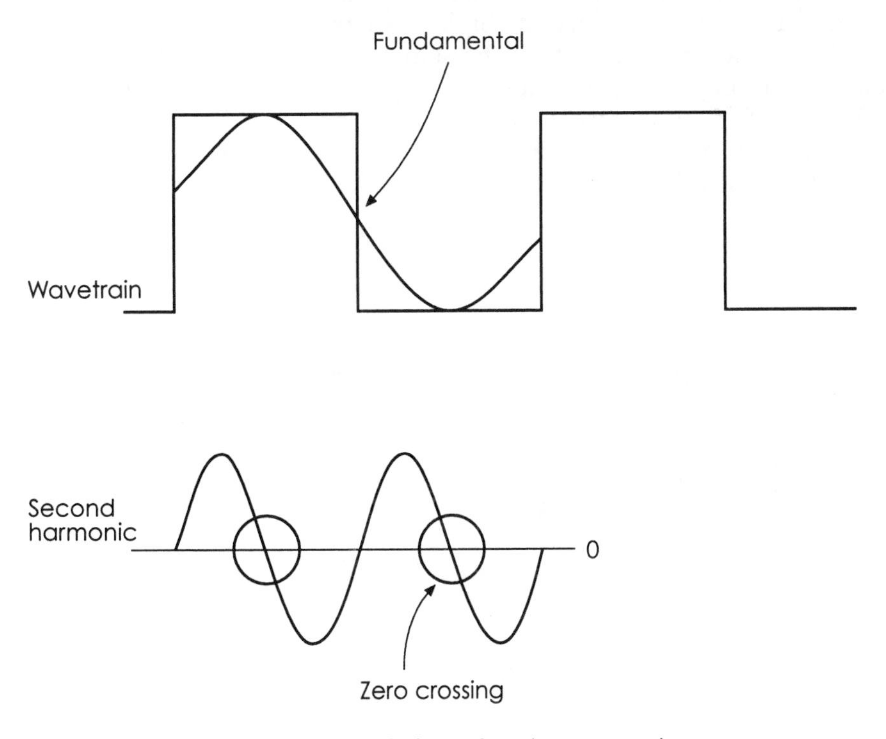

Figure 2.6 Sample timing - Synchronous systems

Networking systems should not prescribe what can and cannot be signalled. They should simply act as a transport medium for any signal and this means that 10101 and so on could legitimately appear in a message. What is the problem with this? Consider the ideal stated earlier: a sample is best taken in the middle of a bit period. Figure 2.6 shows the 10101 wave together with the second harmonic of the fundamental. It can be seen that the positive to negative zero crossing of the second harmonic marks precisely the ideal sampling instant. Regrettably, the equal mark-to-space ratio or symmetry exhibited by the wave train results in no second harmonic being present, only odd harmonics. Non symmetrical signals such as 11010011110 do contain second harmonic content and this component could be filtered out to provide a sampling trigger signal from an edge detection circuit. The system diagram for such a design proposal is given in Figure 2.7. The problem is that, during times when the signalling has a sub sequence such as 101010, the sampling trigger would be lost.

How can the proposal be turned into a reliable system that is insensitive to symmetry? The answer lies in altering the form of binary signalling to ensure permanent non symmetry. One way of doing this is 'half bit time 1' signalling. Another option is a more sophisticated scheme such as 'biphase mark'. These two schemes are illustrated as Figures 2.8(a) and (b) respectively. The rule for biphase mark schemes is that the transmitter causes a transition at the beginning of every bit period but generates a second transition in the middle of the bit period only when a '1' is being signalled. If a '0' is required there is no second transition. To decode biphase mark schemes at the receiver, two samples are taken for each bit. The sampling triggers are derived from the fourth harmonic content. The two samples are applied as inputs to an exclusive OR function which provides the original transmitted bit state. Figure 2.9 illustrates the point. There are many other binary modulation forms, some of which are shown in Figure 2.10.

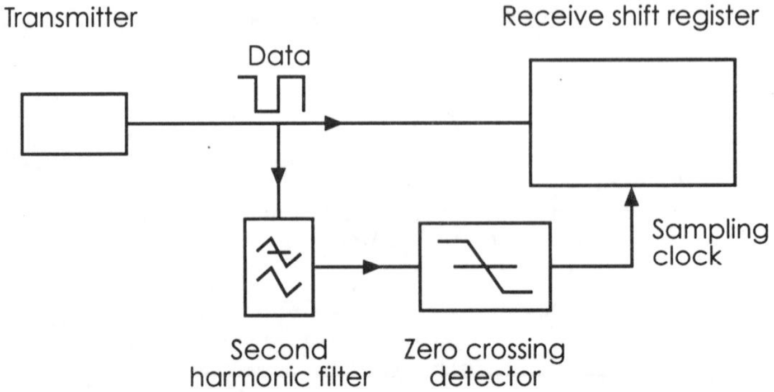

Transmitter

Receive shift register

Data

Sampling clock

Second harmonic filter

Zero crossing detector

Figure 2.7 Synchronous sampling system

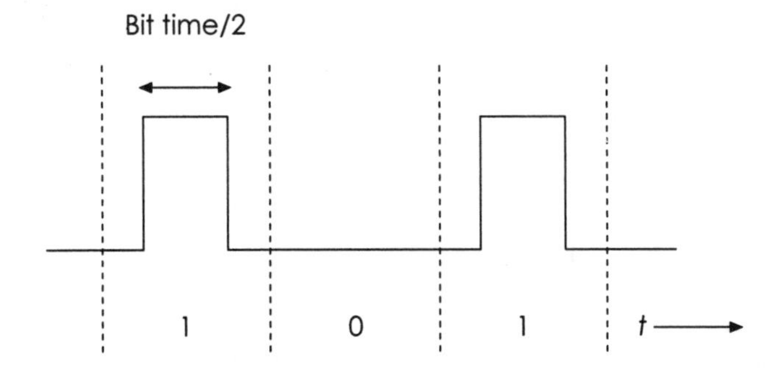

Figure 2.8(a) Half bit time "1"

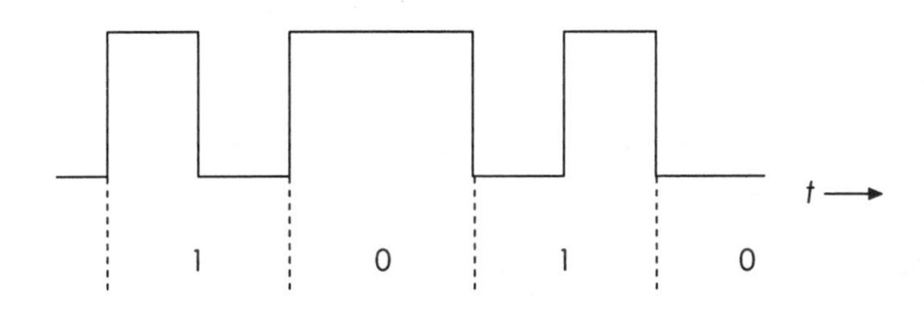

Figure 2.8(b) Biphase mark

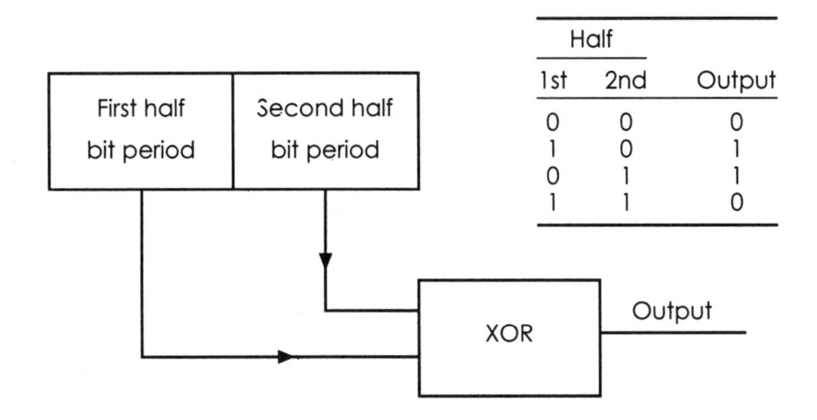

Half 1st	Half 2nd	Output
0	0	0
1	0	1
0	1	1
1	1	0

Figure 2.9 Decoding biphase mark

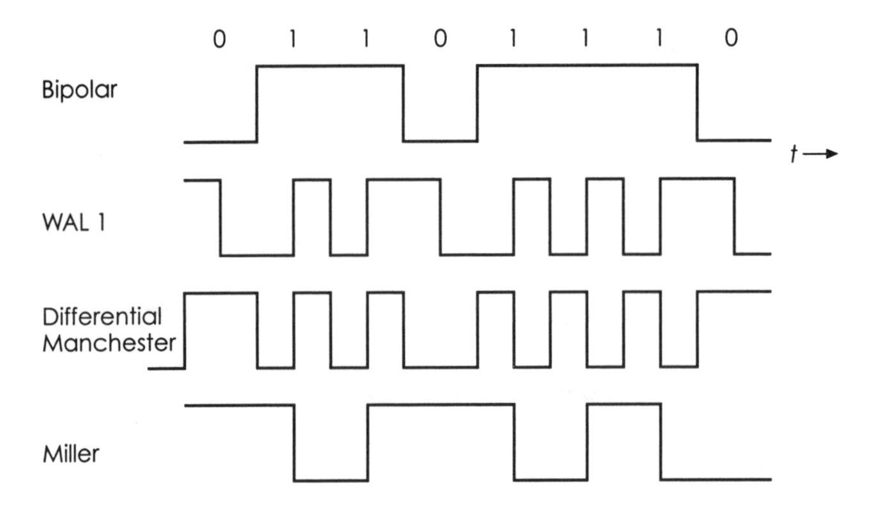

Figure 2.10 Other codes

2.2.2 Data Blocks

Since synchronous signalling derives its sampling triggers from the bit stream and does not need start and stop bits, there is no need to send each entity separately. ASCII characters could be sent end on end. Inevitably, there would be times when there was no data to transmit and if this were so data transfers could be visualised as the movement of blocks of characters. Once you look at the issue in these terms it is clear that the data blocks might benefit from having structure.

Imagine a receiver dealing with a string of 0s and 1s. Bit sampling derived from harmonic content cannot begin until the bit stream does. Even then there will be a short delay until the triggers appear and in that time the first bit or two will have been missed. From then on the receiver will not 'see' the bits in correct registration as it tries to interpret 8 bit characters. The effect is demonstrated by Figure 2.11 and is a powerful reason for 'run in' coding at

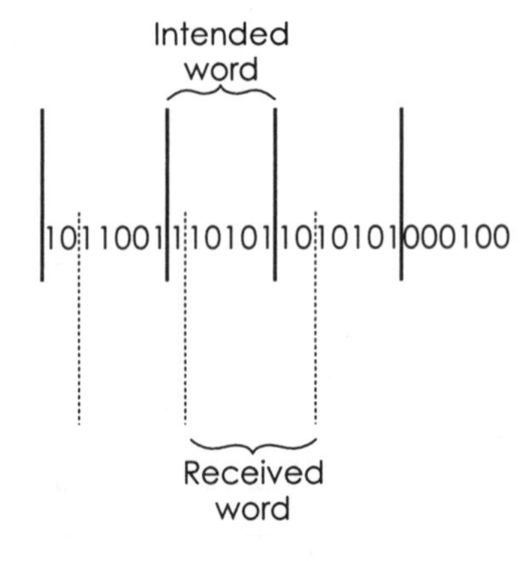

Figure 2.11 The need for run in codes

the front of the data block. Very often a special character, 'SYN' (ASCII–16 hexadecimal) is used for this and, because the first few bits are likely to be lost, two or three 'SYN' characters will act as a run in code. Their presence allows the receiver to lock on to the correct 8 bits making up each subsequent character. A typical system will employ a binary comparator to detect the SYN character. When sampled bits are received they are clocked into a shift register. The system performs a bit-wise comparison and only when the word in the shift register is the same as the word in a buffer register is the output from the comparator activated. The comparator output is used as a trigger which identifies the correct registration of signalled bytes. The suggested system is illustrated by Figure 2.12.

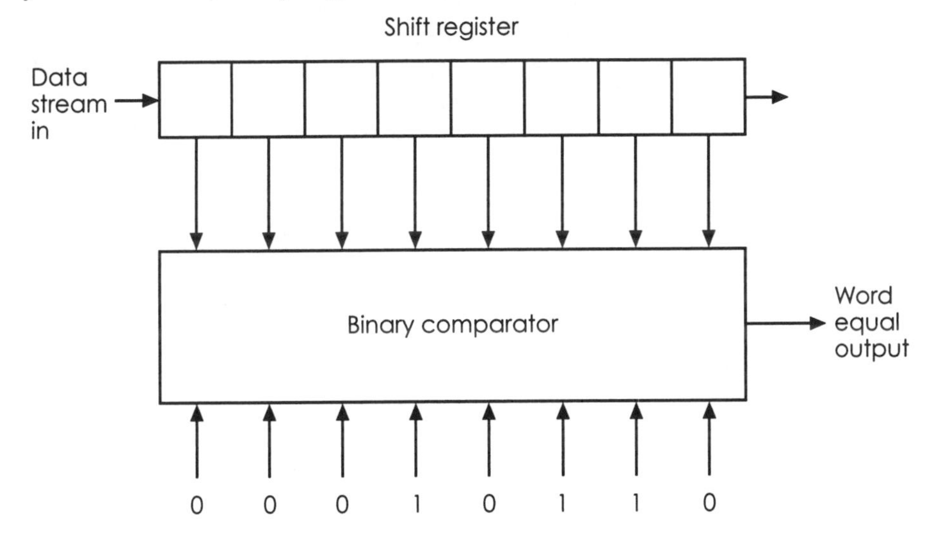

Figure 2.12 Binary comparator

2.2.3 Bit oriented signalling

Although the principles of synchronous systems have been explained in terms of byte oriented data, there are some forms of data where it is not particularly convenient to operate in that way. It may also be a constraint that data frames must contain a whole number of bytes. Byte oriented systems reserve some 0 and 1 patterns for control purposes, e.g. SYN (00010110). This may be very restrictive and to understand the problems and constraints, imagine the signalling of the pixel (picture element) states making up a graphics screen. Clearly, it is unhelpful to say that a particular pattern, which is part of the picture, may not be sent because it could be mistaken for a control pattern such as SYN.

To deal with these problems an alternative data frame structure is to used to allow data frames to contain any number of bits with any pattern. The main objective is to obtain data transparency, a term which means that any necessary control patterns used may also be allowed as genuine data. Generally, only one control pattern is necessary and it is used as a delimiter to mark the start and finish of the frame. Typically, this delimiter or opening/closing flag is 01111110. The question is: how can this pattern also be allowed to appear as a bit pattern in the message?

2.2.4 Bit stuffing

Whilst each frame commences with the opening flag pattern, 01111110, once the pattern has gone the transmitter monitors the subsequent data looking for five consecutive 1s. When five 1s are spotted, the transmitter inserts a 0 bit state to follow. The outgoing bit stream for a genuine 01111110 is changed to 011111010. To end the frame the closing flag pattern 01111110 is created. In this way the opening/closing flag pattern only occurs unaltered at the beginning and end of a frame, where it acts as a delimiter. Where it would have appeared within the frame it is modified by the insert 0 policy. The receiver simply strips off the opening and closing flags and applies the inverse algorithm which looks for five consecutive 1s and deletes the next 0. In this way, even 01111110 may be carried (disguised as 01111101). No other pattern is illegal and may be sent unaltered. The overall process is called 'bit stuffing' and because a frame can consist of any number of bits the 'extras' are no problem at all. Figure 2.13 shows some original data, how it appears after bit stuffing, and how it is constituted after recovery.

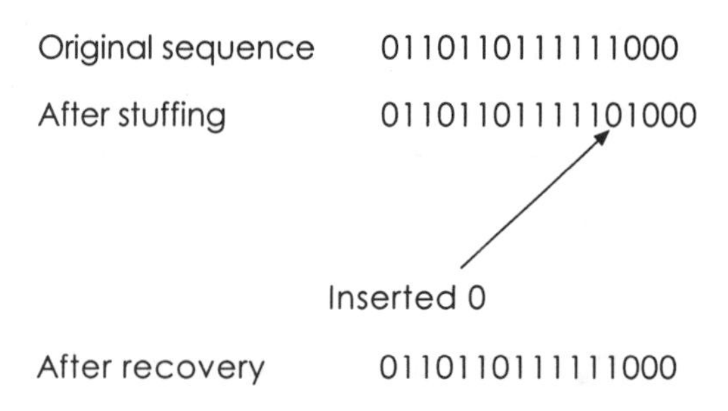

Figure 2.13 Bit stuffing

2.3 Baseband signalling

It has been shown that square wave trains may be analysed in order to determine their fundamental frequency and harmonic components. To carry a signalled wave satisfactorily it is necessary for the transmission media to possess sufficient bandwidth to accommodate significant numbers of harmonics. This is best understood by example. Consider a 10 megabit/second system. Each bit is sustained for 0.1 microseconds and since the period of the fundamental is two bit periods, its frequency will be 5 MHz (frequency = 1/period). Allowing that every harmonic up to the seventh must be catered for, the signal will have a bandwidth of about 35 MHz and the associated spectrum will be as shown in Figure 2.14.

Such a signalling strategy is normal for systems that do not operate in an electrically harsh environment. When network cabling shares trunking and conduit with power cables or when highly inductive machines and sparking

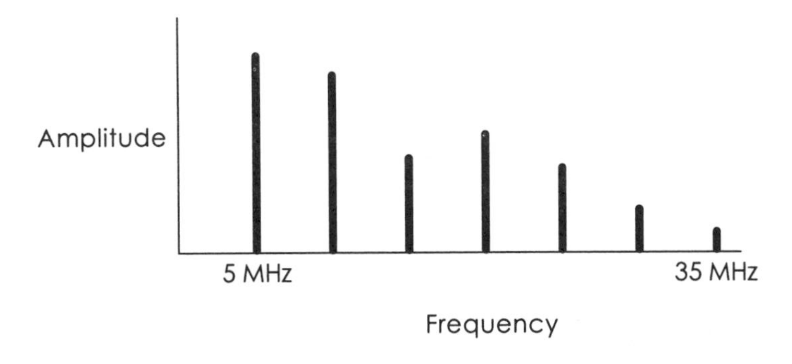

Figure 2.14 Harmonic content of 10Mb/s signal

contactors, chopper power supplies and motor speed controls are present, such as in a factory, the story is different. All these factors create electromagnetic noise which will cause significant interference to baseband signals. Fortunately, the characteristics of industrial noise are well known and it is generally described as pink noise with a roll-off around 300 MHz. A Bode plot for industrial interference and its relationship to a baseband signal is given in Figure 2.15. The problem is that the signal to-noise-ratio may be poor and interference may cause large numbers of corrupted data bits in the message.

2.4 Carrier band signalling

Generally, a carrier is a higher frequency entity than the message signal. A special circuit, called a frequency changer or mixer is used to superimpose the message onto the carrier. In this way the message is translated up the frequency spectrum to a convenient place. With an industrial environment the carrier may be chosen to be above the pink noise at around 650 MHz. Figure 2.16 shows the effect of this process. Since the message frequency and noise components are now separated, filters may be used to eliminate the noise. The use of carrier band methods makes signalling relatively free of corruption. The principles underlying the frequency changer are described in

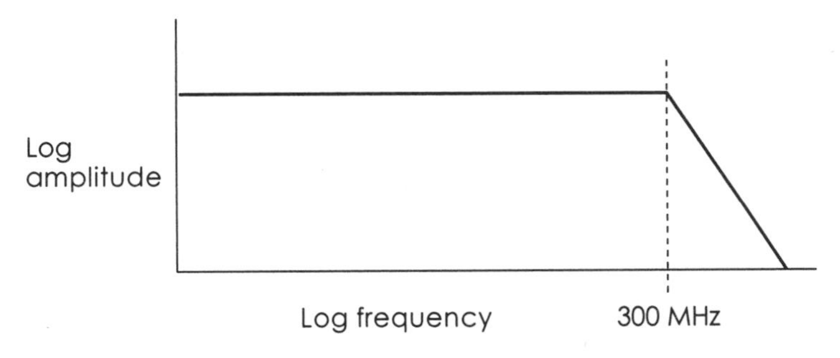

Figure 2.15 Industrial interference spectrum

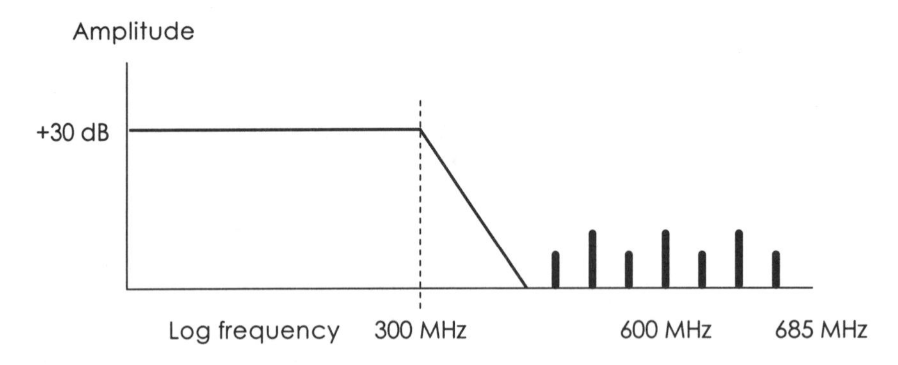

Figure 2.16 Carrier band spectrum

Appendix A. The concept of filter technology is dealt with in Appendix B. Naturally, the use of a carrier band system requires that the transmission medium has sufficient bandwidth. If all of the baseband frequency components are accommodated within the range up to 35 MHz, for the carrier band version the highest frequency component is 685 MHz. The message must be recovered from the carrier at the receiving end by a circuit that is able to detect the difference frequencies between the carrier and the message components.

2.5 Broadband systems

An extension of the modulated carrier principle allows the available bandwidth of the transmission medium to contain a number of carriers. A message channel may be associated with each carrier so that a common data highway will carry a number of independent dialogues. Each of the multiple channels occupies only a small portion of the overall available bandwidth. Each channel requires the use of detectors or demodulators and bandpass filters. This shared frequency spectrum concept is termed frequency division multiplexing. Figure 2.17 shows how the bandwidth of a cable is used in broadband systems. The expensive nature of broadband provision has made

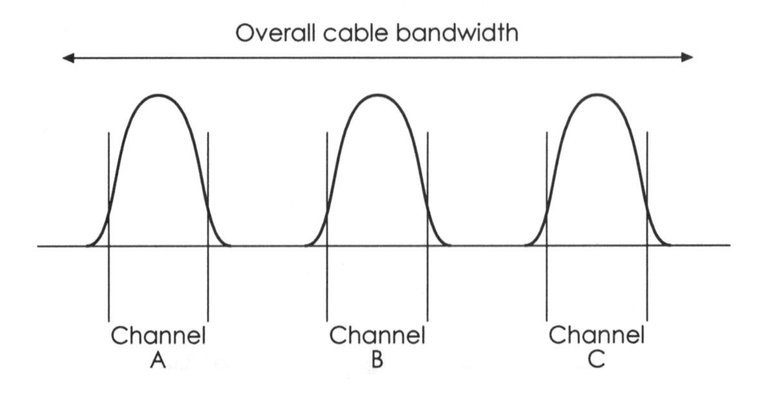

Figure 2.17 Frequency division multiplexing

it unpopular for networks carrying purely alphanumeric data or executable code downloaded in a client-server relationship. Broadband scores when integrated services such as videophone or speech are to be carried alongside normal programs or data. This is because the media can carry simultaneous dialogues of differing kinds. The normal channel width is about 6 MHz, which is sufficient for video but which will not allow signalling at 10 Mb/s. To achieve 10 Mb/s it is necessary to concatenate adjacent channels.

One of the architectural features of broadband systems is the use of 'head end' units. The arrangement is shown by Figure 2.18. The head end unit provides a frequency translation function using frequency changer or mixer principles together with filters. Most systems are uni-directional for a given frequency. The overall bandwidth is split in two and the lower range is generally used for node transmit channels. The upper half of the range serves as a series of receive channels. Each node transmits on a transmit channel which is not heard directly by the intended receiving node. The transmission is heard only by the head end unit which translates the data onto a different channel in the receive range and retransmits the data. In this way the intended receiving node gets the data via the head end unit. In the simplest model each node has a fixed receive channel and a fixed transmit channel. When a link is requested, the head end unit acts as an intelligent switch which performs the appropriate frequency translation for a given link. The head end unit is a control centre. The frequency spectrum for the channels includes a 'separation' or 'guardband' between the channels to ensure that the filters in the head end and in the nodes can effectively discriminate between channels and avoid any interference.

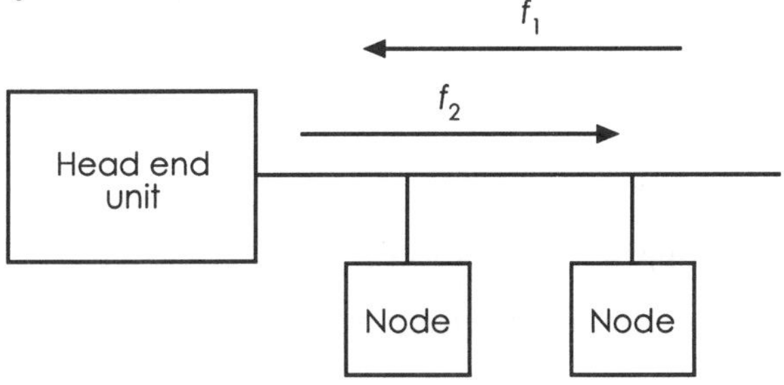

Figure 2.18 Broadband architecture

2.6 Case study decisions

Appropriate choices for new networks must take into account the requirement specification. The case study in this book has a first objective of networking an office environment. The distances involved, four floors of 200 metres end to end, are well within the scope of local area networks. Bearing in mind the constraint that the ISO OSI 7 layer model is to be used in the design, it is necessary to prescribe layer 1, the physical layer.

Much of what has been discussed in this chapter is appropriate to this first task. Since the environment cannot be considered as harsh in electromagnetic terms, the choice of a baseband system using coaxial cable seems natural. If it is assumed that the network can be implemented by an unbroken run of cable then the maximum signalling range within the headquarters building will not be more than 400–500 metres. Accordingly, the signalling levels may be adequately set at – 12 to + 12 volts. The requirement for a maximum of 100 connected computers per floor will affect the design of the line driver circuits for the node connections. Important parameters here will be: a driver output impedence sufficiently low to allow the node to source sufficient current to cope with the parallel input impedence of node receiver circuits; and provision of a sufficiently high slew rate to allow signalling at 4 Mb/s or better.

The requirement specification clearly expects the network to handle heavy traffic and because of this the extra efficiency of synchronous signalling would suggest its use instead of an asynchronous option. This being so, which of the possible forms of binary modulation should be used? There is little to choose between any of the possibilities illustrated by Figures 2.8 and 2.10. Since WAL encoding is used by national data carriers and it also appears as an option for many network monitoring systems it would be an appropriate choice on grounds of compatibility.

The question of structure in the data blocks is an issue more properly discussed under ISO layer 2, the data link layer, but the boundaries between the layers are often blurred and simple elements of structure will be discussed here. It is not clear from the requirement specification whether traffic will consist exclusively of purely ASCII characters, but the improving human computer interface seems to demand ever more sophisticated graphics effects. It would be unsafe to expect a modern system to carry ASCII only and it would seem sensible to prescribe a system which provided 100 percent data transparency. This being so, the choice is a bit oriented system.

Questions

1 What is meant by 'digital modulation'?

2 Explain the role of digital modulation in synchronous data transmission.

3 Differentiate between asynchronous and synchronous signalling and show the appropriateness of each to specific cases such as:-

(i) A high traffic density archival link between main processing centres.

(ii) Local communication between a dumb terminal and a computer.

4 Explain how carrier band signalling is helpful when the environment that a network is operating in is harsh electromagnetically.

5 Describe the purpose of guardbands in the spectrum used for broadband systems.

3

Single network issues

3.1 Interfacing to the network

Connection to a network requires an interface. The network is likely to be operating with synchronous transmission at relatively high speed, usually between 1 and 20 Mb/s. Terminals, host computers and PCs are not usually provided with appropriate communications ports and for this reason alone the need for an interface is clear. There are other reasons. Although PCs and terminals are general purpose machines the wide diversity in the detailed working of different networks would be a nightmare for system providers. It is better for commercially available networks to provide specific software for their own case, working through one of a few standard interface circuits. Whether the interface circuit is a plug in card for a PC bus or a stand alone box it will be convenient to separate the functionality of the interface from that of its associated peripheral or computer. In this text an interface will be referred to as a network 'node'. It will be assumed that the peripheral device or computer passes messages to the node, which is then responsible for launching the message onto the network and delivering it successfully. Received messages will be decoded from the specific form demanded by network transmission and handed to the peripheral. Two peripherals communicating over the network will not be aware of its presence. Its action will be transparent. The way in which a node is connected to the network data highway is important to the detailed action of the system, whether the highway be a cable,optical fibre, or radio link. There are two main alternatives, 'point to point' or 'broadcast'.

3.2 Point to point links

The characteristic of the point to point form of node connection is that the network messages are 'internalised' by the node. In other words, the node takes information in and stores it, perhaps only momentarily, before sending the message out to the next node in the network. The network could be considered to be a series of end to end links joined together. The node circuitry is in series with the data highway. Figure 3.1 illustrates the concept.

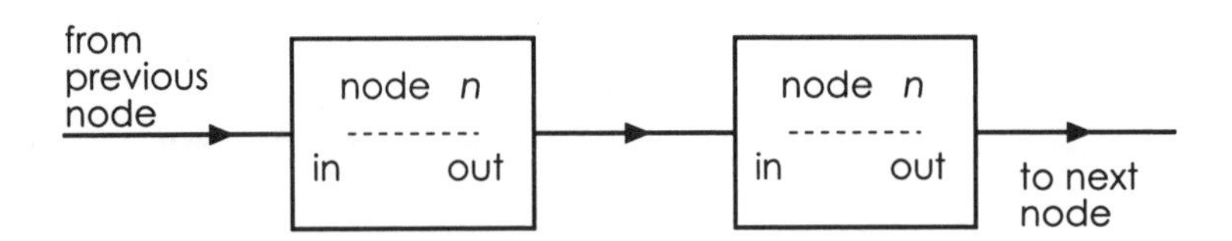

Figure 3.1 Point to Point links

Point to point designs are more traditionally associated with long distance networks using links provided by external providers but some local area networks have been implemented this way and so the necessary design background will be discussed. The 'pass through' nature of data interraction with nodes has serious consequences for the network because the failure of a single node can result in parts of the network being unreachable. Designers would minimise the risk by creating alternative pathways to eliminate such problems. This inevitably requires an understanding of routing, delay analysis and queueing theory. Inside point to point nodes data buffering must be provided. The number of data buffer memories, their size and their allocation would be critical decisions affecting how well the network handled congestion and flow control. The most serious problem is how to make the network invulnerable to deadlock. To appreciate the danger, imagine two nodes A and B. Each has 6 buffers. All 6 buffers in node A are full with messages for B. All 6 buffers in B are full with messages for A. Neither node can accept any data because their buffers are full. They are said to be deadlocked. The magnitude of the design task is clearly very great and rather than pursue design solutions an alternative connection strategy will be examined.

3.3 Broadcast links

With broadcast links, nodes are connected to the data highway as 'taps'. This means that the nodes are effectively connected in parallel with each other. Figure 3.2 shows the configuration. Data passing over the data highway is available to all nodes. In this way data may be addressed to one, any number or all the nodes. It is this last option which gives rise to the name 'broadcast' for the connection mode.

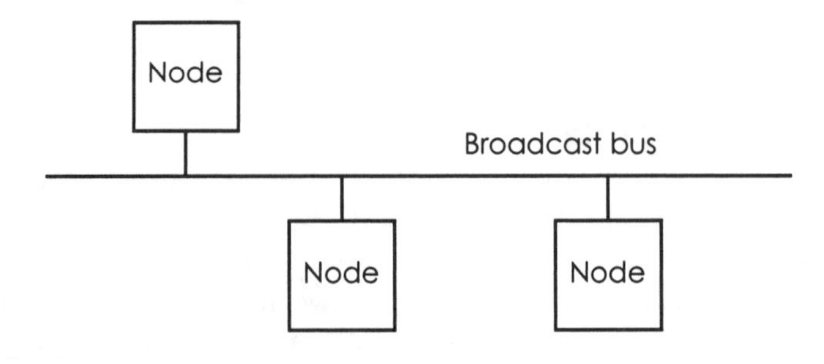

Figure 3.2 Broadcast links

This connection scheme has proved the most popular for local area networks because a lot of the problems associated with point to point design are not relevant. There are problems to solve in implementing broadcast systems but they are more concerned with topology and access method. Topology is shape or layout and it becomes particularly significant because a single highway must reach each peripheral or computer irrespective of its geographical position. The access method is important because it defines the way in which nodes are allowed to launch data onto the data highway. This must be done in such away that the highway is used to a level of efficiency appropriate to the network traffic demand. These two areas must be examined in detail. Topology is an ISO physical layer issue whilst the access method is a part of the data link area of interest.

3.4 Topology

3.4.1 Bus

Figure 3.3 shows the general arrangement. A highway is simply a length of cable (or other medium) with its ends terminated by impedence matching devices. These devices are crucial to the correct operation of the network because without them the highway terminates in an open circuit which will cause signal to be reflected back from the terminations. There would then be interractions between signals originating from a node and those reflecting from the ends. The resulting standing waves would render the network unusable. Nodes are tapped onto the highway in the manner shown in Figure 3.2.

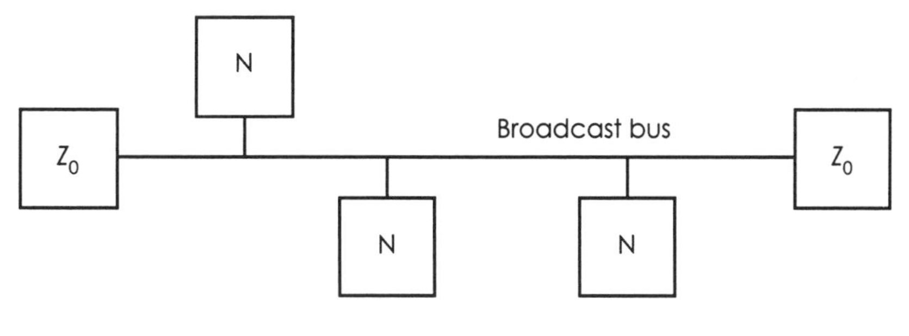

Figure 3.3 Bus topology

When a node transmits data the message propagates outward from the node in both directions and eventually dissipates in the terminations (see Figure 3.4). Data signals travel through the cable as guided electromagnetic waves and their velocity is usually about half the speed of light, say about 150 million metres per second. If the highway were 1000 metres long and a node at one extreme end sent a message to a node at the other the receiver would first see the message about 7 microseconds later. The fact that communication is not instantaneous will be an important feature in later access discussions.

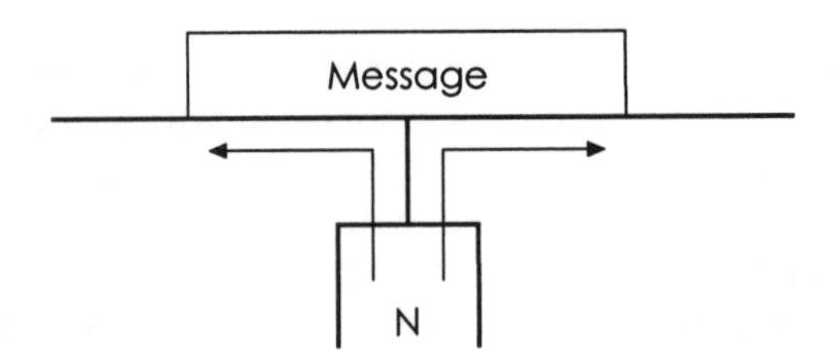

Figure 3.4 Bus broadcast transmissions

3.4.2 Ring

If the data highway forms a closed figure then the topology is defined as a ring. With this structure, data launched onto the highway must be removed at some time, otherwise it will circulate indefinitely. Normally the node originating the data is charged with removing it subsequently, one hopes after the receiving node has read the message. Figures 3.5(a),(b) and (c) are all examples of a ring.

Ring nodes contain a line driver or amplifier and a tapping point to enable the reading and launching of data. The presence of the driver implies that data may travel around the ring unidirectionally (amplifiers are one way devices).

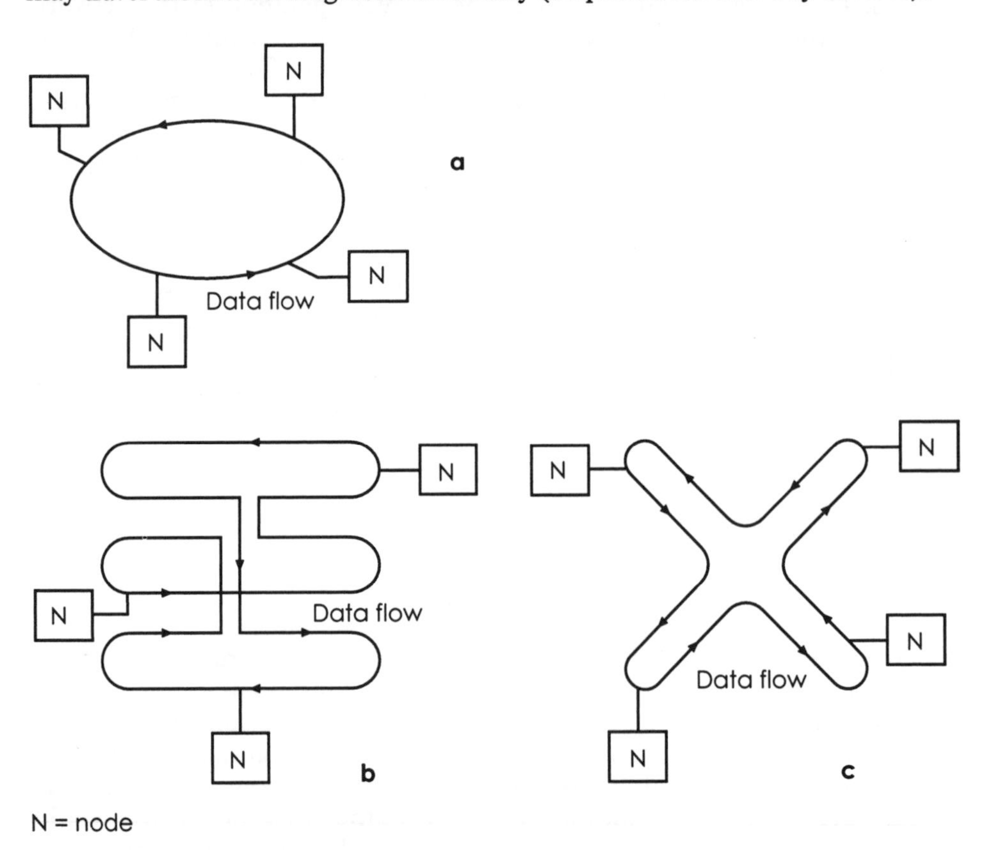

Figure 3.5 Three ring variants

3.5 Access methods

One of the greatest differences between networks is the way that they handle the launching of data onto the highway. Particular access methods are often portrayed as 'locked' to particular topologies. This is a misleading byproduct of the fact that the two most commercially successful network architectures use dissimilar layouts and dissimilar access methods. In truth, any access method can be used with any topology.

The access methods available must solve the important problem of arbitrating or avoiding data collisions caused by multiple simultaneous transmission and subsequent data corruption. In general, the problem is tackled either by not allowing multiple simultaneous transmission, or, if this is allowed, providing a detection and recovery strategy. In the following detailed discussion it will necessary, from time to time, to imagine a network in operation. To assist with this, one topology will be used for all described access methods. Later, the topology will be changed and implementation of the access methods to the new layout will be discussed. In this way, an emphasis is placed on the freedom of the designer to mate any topology with any access method. The initial topology used will be the bus.

Access methods may be categorised into those that allow the specification of a 'worst case' delivery time for data (deterministic) and those that do not (probabalistic). The application to which the network is put will strongly affect the degree to which determinism is critical. Safety legislation and common sense might disallow probabalism in a network used for control and shutdown of a nuclear reactor, whereas in an office environment a few milliseconds unexpected delay will hardly be noticed.

3.5.1 Contention

In its original form and in commonly used later derivatives, contention is a purely probabalistic system. As such, contention is a synonym for collision because this access method expects collisions to occur and provides a recovery strategy, albeit subject to variable and undefined delay. The example used to explain the concept is called CSMA/CD (carrier sense, multiple access, collision detection). A node wishing to send a message over the highway may do so at any time, subject to a few provisos.

First, the node must 'listen first'. This is to ensure that no other node is in mid-message. There is no sense in allowing avoidable clashes to take place. If the node 'hears' nothing it may assume that the highway is idle and may simply transmit its data frame. If another node is transmitting the node must politely wait for the highway to become idle. The perfect feedback provided by a broadcast bus must be used by a transmitting node to monitor its own transmissions.

This polite conduct might be thought to ensure that collisions do not occur. This is not so: you will remember that messages take a finite time to propagate along the length of the bus. Imagine a bus where two nodes have data to transmit. They are sited at opposite ends of the bus and because of this when one transmits, the other will not hear the transmission for a time equivalent to the one-way travel time of the line. In Section 3.4.1 an example quantified the delay at about 7 microseconds for a given case. Although one node transmits, another may transmit after the first but before hearing the other. There is a window of opportunity for collision.

Because the nodes are required to listen to their own transmissions data corruptions will be detected and CSMA/CD prescribes that on detecting corruption a transmitting node should reinforce the event by sending a jamming signal. In this way there is no ambiguity and a collision will be clearly recognised. The response to data collisions is that after jamming the highway all transmitting nodes stop transmitting and wait a pseudo random delay before trying again. This is called a back off strategy. If, on the second try a collision again occurs, a node will back off and double its delay timeout. This creates a binary exponential back off. Figure 3.6 demonstrates the action.

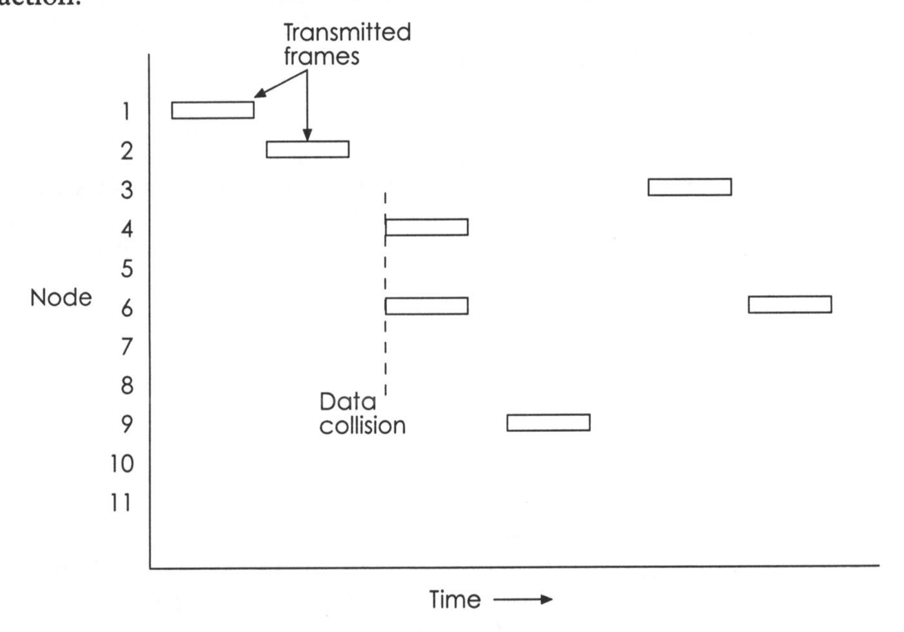

Figure 3.6 CSMA/CD collision

In light traffic conditions there are unlikely to be many data collisions and the vast majority of messages will get through with no delay. As the traffic offered to the network increases so also will the number of collisions and subsequent retransmission delays. It is possible to imagine that at the limit all nodes will wish to transmit and collisions will occur continuously. If this happens the network will become jammed solid with permanent contention. The throughput will fall to zero and the network is in a failure state from which it cannot recover unaided.

Some assumptions have been made in describing this scenario. For instance, it has been assumed that potential sending nodes listen continuously, ready to jump in as soon as the the highway seems free. This style of operation is called 1-persistant CSMA/CD. If a variable G is defined as offered load in terms of attempts to send a frame whenever the highway is free then the utilisation figure (≤ 100 per cent) is given by :

$$S = (Ge^{-G}(1 + G))/G + e^{-G}$$

where S is the utilisation.

Obviously the best utilisation will occur when one frame attempt is made each time the channel is free. Any less and the highway will be idle for a proportion of the time, any more and collisions will detract from the throughput. Figure 3.7 shows the plot of S against G. As might be derived from the equation and from common sense, as G tends to infinity, the utilisation and hence the throughput tend to zero.

Many decisions in engineering involve a trade off. It has been found that non-persistent CSMA can improve the highway utilisation but at the cost of greater average delays. Non persistence simply means that nodes do not monitor continuously, ready to pounce. Instead, they apply a random delay between even looking to see if the highway is clear. Figure 3.7 also demonstrates the results of this.

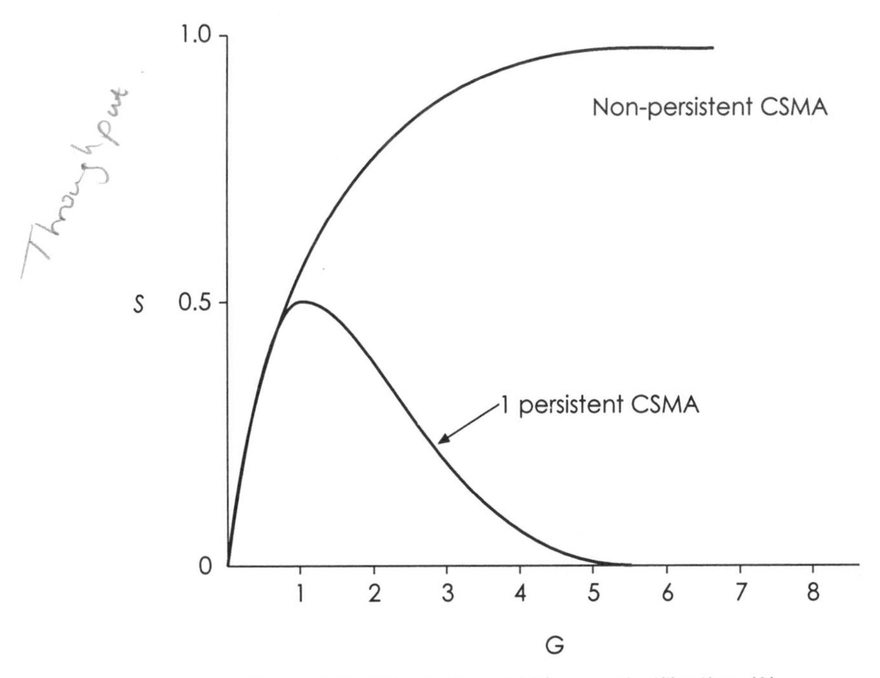

Figure 3.7 Offered load (G) vs net utilisation (S)

3.5.2 Slotted contention

The relatively unsatisfactory elements of the contention systems already described led workers to try to address the problems of poor throughput. The first step was to compartmentalise time. Time is split into periods when nodes contend or compete to send their frames and times when data frames themselves are sent. These basic periods alternate. The arrangement is demonstrated by Figure 3.8. There is an instant improvement caused by the fact that collision vulnerable time is reduced by a large factor.

Design decisions critical to this strategy concern the length of time

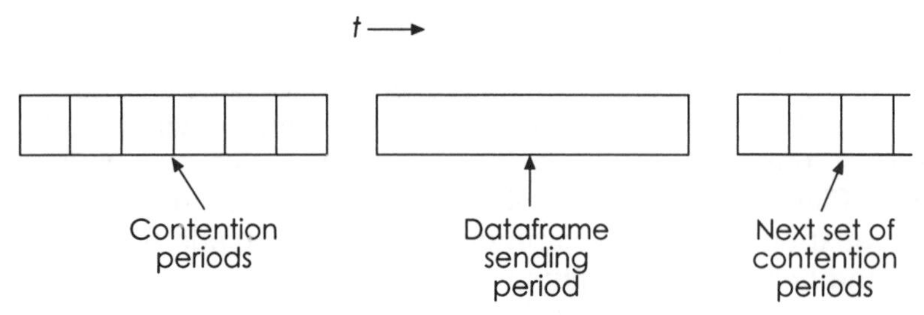

Figure 3.8 A slotted contention scheme

allocated to contention slots and the way slots are used. A node wishing to transmit waits for the next contention period and after choosing a particular slot transmits a special signal. There are many options for the nature of this signal but the important feature of it is that nodes must be able to detect multiple transmissions. The question is: how long must the node transmit in the slot to be sure that no other node is contending? The answer is twice the maximum propagation delay on the highway between the furthest nodes. The logic of this is as follows. A node may transmit and just as its signal reaches the farthest node, that node transmits. The original will have to wait twice the one way travel time of the highway in order to detect this worst case situation.

It makes sense to tailor the length of a contention slot to suit this underlying timing since, if contention slots are unnecessarily long efficiency will suffer and if they are too short a collision may go undetected. On a 500 metre highway the slots will be about 5 to 7 microseconds long. As is shown by Figure 3.8, contention slots are repeated until a node contends successfully, after which the node may send its data frame. Usually, a maximum number of uncontended slots is allowed after which, if no claims are made, the highway enters an idle state. The highway may therefore be in one of three states: contention;data transfer; or idle.

One of the key strategies to be decided is how quickly a node should try again after a collision in a contention slot. 1-persistency would not be helpful and other probabilities have been tried with success. 0.1-persistent means that a node with traffic to send will transmit on a contention slot boundary with a probability of 0.1. Such a strategy considerably reduces collisions but does introduce a delay. If the probability is P, then the average delay for this method is :-

$(1 - P)/P$ slot times.

For 0.1 persistence the average delay is therefore nine slots but the best utilisation of the highway can be very high, in the region of 80–90 per cent. Even so, the utilisation still falls when many nodes are contending, falling to about 70 per cent when there are six or seven attempts per slot. The generic term for this type of approach in slotted systems is P-persistant.

The slotted contention mechanism that has been described above is that of the successful commercial network marketed as Ethernet. So far, nothing has been said about how contention networks may arrange for positive acknowledgment of data frames. In these systems transmitted frames never return to their originating nodes and they are read 'on the fly' by the receiving nodes. Because of this is it necessary for receivers to generate discrete data frames that are addressed to the sender and which carry the message 'data frame acknowledged'. For slotted systems the 'stop and wait' protocol discussed in Chapter 4 can be made reasonably efficient. The way that this is done is that the first contention slot period following a data frame transmission is reserved for the use of a receiver. If the receiver transmits its special contention signal in that period, then the transmitter interprets this as an acknowledgement. It should be made clear that no other nodes are permitted to use that particular slot and so its use is unambiguous. In Ethernet, collisions do not occur when data frames are being transmitted. They can and do occur during contention slot periods. Any contention adversely affects throughput and collision free methods have been devised. The Ethernet data frame format is shown in Figure 3.9.

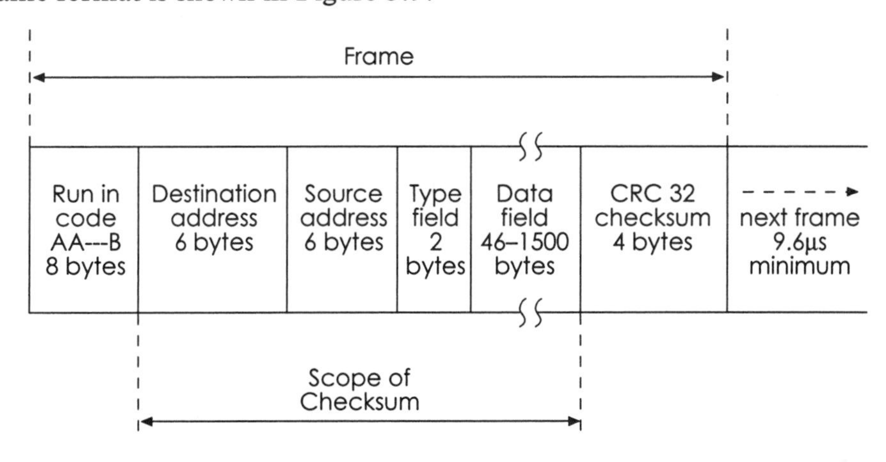

Figure 3.9 Ethernet frame

Because under Ethernet the highway can be idle it is necessary for data frames to ensure that receiving nodes have time to bit synchronise. This is achieved by the use of a preamble to the frame. The preamble consists of eight bytes, the first seven of which are the pattern 10101010. The last is 10101011. The alternating nature of 0s and 1s provides ample harmonic content for the receiver to lock to. The preamble is followed by source and destination node addresses and a 'type' field. The type field is used to differentiate between frames that are for control purposes e.g. an acknowledgement, and pure data frames. Additionally, the type field carries the frame number and a code identifying the format of the data in the data field. The data field itself can be any length between limits of 46 and 1500 bytes. Error detection is provided by a cyclic redundancy check word of 32 bits, based on a polynomial of

104C11DB6 hexadecimal. Cyclic redundancy checks are discussed in Chapter 4. An interesting feature of the Ethernet system is the time allowed between data frames to allow receiving nodes to absorb and check the data. Consecutive frames have a minimum spacing of 9.6 microseconds.

An Ethernet bus is 500 metres long when tapped, although an untapped length of 1000 metres is permitted. The strange notion of an untapped highway will be explored in Chapter 5 which looks at the interconnection of separate highways. Figure 3.10(a) shows the basic parameters. A cheaper alternative still signalling at 10 Mb/s but allowing a maximum length of 185 metres is illustrated in Figure 3.10(b). These two options are referred to as 10base5 and 10base2 respectively.

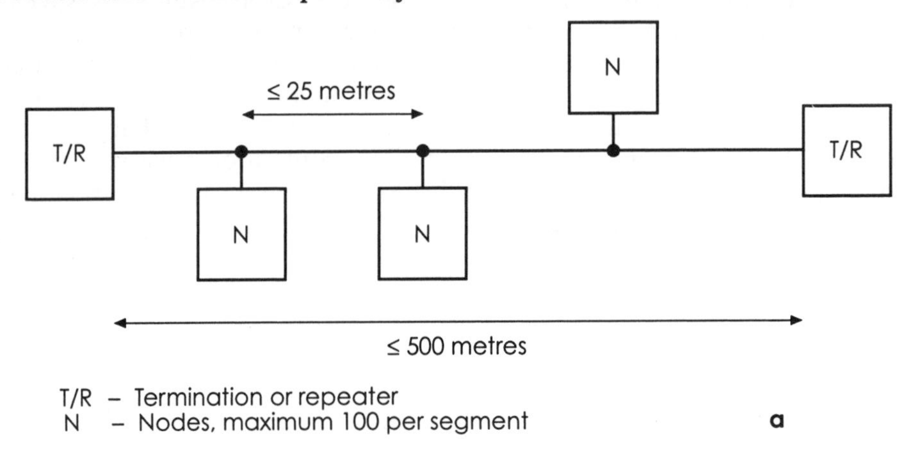

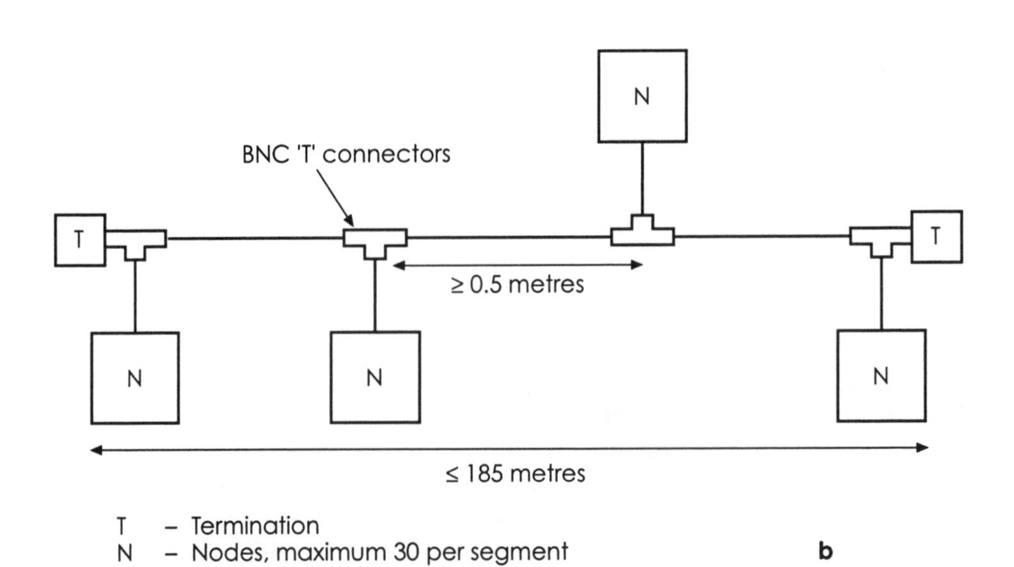

Figure 3.10 Ethernet segment (a), 10base5, and (b), 10base2

3.5.3 Bit mapped slots and derivatives

The bit mapped slot method retains the notion of discrete contention and data transfer periods. It disposes of the collisions that otherwise could occur during contention by allocating a specific slot to each node on the highway. Only node number one can transmit in slot one, and only node seven in slot seven. At the end of the contention period, each node will know which nodes wish to transmit data frames. Perhaps it is one, three and six. All nodes know the situation and one, three and six transmit their frames in strict order. After the last has transmitted its frame a new contention period begins. The obvious disadvantage of this method is that the number of slots may be very large and to offset this problem each slot may only be the time necessary for a single bit. A reservation may be made simply by each node having a frame queued for sending asserting '1' during its own contention slot. When no traffic is passed the highway carries continuous bit maps of slots. If there are N slots in a bit map cycle, on average a node will have to wait $N/2$ slot times in order to make a reservation. It was soon spotted that certain inequalities were present in this scheme. A low numbered node might just miss its slot. It then has to wait for the rest of the slots to pass before a new slot cycle begins and it can make a reservation. Once the reservation is made it must then wait for that slot cycle to end before it can send its frame. A high numbered node might just miss its slot. It waits much less time in the remaining cycle before a new cycle begins and it may reserve a slot. It has been calculated that low number nodes must wait, on average $1.5N$ slots whilst a high number node waits on average only $0.5N$.

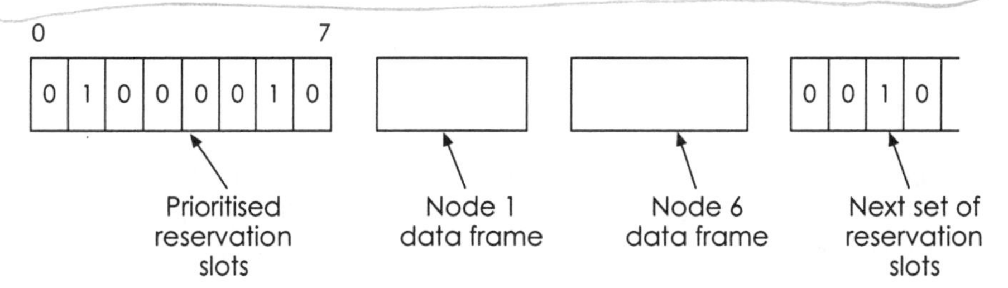

Figure 3.11(a) Bit mapped slots

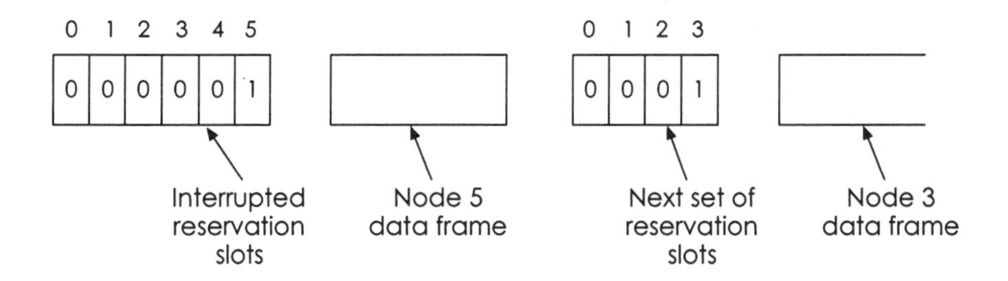

Figure 3.11(b) BRAP action

A derivative called broadcast recognition with alternating priorities BRAP attempts to do something about the inequalities by inverting the bit map on alternate cycles. Another difference is that the bit map slots are aborted as soon as a node reserves a slot. The node then transmits its data frame and the bit map slots are resumed. These collision free systems are illustrated by Figure 3.11(a) and (b). In these collision free systems a worst case delivery time can be guaranteed and so the systems are deterministic.

It is worth noting that the simplest contention systems yield low delay when traffic offered is also low. The more complex options offset contentions by trading for extra delay. Some network designers have sought the best of both worlds by using systems which switch actions when a specific traffic offered threshold is exceeded, returning to a simple system later when the traffic dies down. This kind of thinking is most important when highway lengths are long. This is because of the increased collision risk inherent in longer two way propagation times.

3.5.4 Token passing

The approach taken for token passing is that a node may only transmit when has permission to do so. The permission takes the form of a 'token' which is passed from node to node cyclically. Some designers view the concept as a novel form of time division multiplexing which dynamically allocates time only to those stations that wish to send. The token passing system is fully deterministic.

When the network is active a special control node takes the responsibility for creating and launching the token, which generally takes the form of a short frame as shown in Figure 3.12(a). Figure 3.12(b) is the full frame format.

The frame control field is the heart of the operation. It contains bits whose function is vital to initialising, joining or leaving the active network. The token frame is passed from node to node in a predetermined order usually based simply on address values. If a node does not want to use the token it increments source and destination addresses and retransmits. If a node does

Run in and delimiter	Frame control	End delimiter

Figure 3.12(a) Short token frame

Run in code > 1 byte	Start delimiter 1 byte	Frame control 1 byte	Destination address 2 – 6 bytes	Source address 2 – 6 bytes	Data < 8182 bytes	CRC 32 4# bytes	End delimiter 1 byte

Figure 3.12(b) Token bus frame

want to use the token, it toggles a bit in the frame control field to indicate the frame is not a token and inserts data into the data field. Once the sending node has sent as many frames as necessary it must relaunch the token, passing it to the next in order.

To make the token passing system work each node knows the address of its preceding and following partners. Problems arise in two main cases. First, what happens if a node or nodes are off line or powered down? Secondly, what happens if a node wishes to leave the system after it has initially been included in the round robin?

The first problem may be tackled by an enquiry sequence immediately after the control node is switched on. The enquiry sequence seeks specific responses from all node addresses, building up a picture of the active nodes. From this activity, nodes will be able to identify their preceding and following partners even if many addresses are missing. So much for initialisation but the population of the network may vary dynamically. A node wishing to leave the network waits until it receives a token and then uses it to send a special frame to its predecessor. The special frame is identified by coded bits in the frame control field and this frame carries the address of the node following the one that is leaving. What if a node wishes to join an existing round robin? Each node occasionally issues a special frame which effectively asks: who succeeds me? The addresses of nodes that may answer are conditionalised by mask bits in the destination address. This enables enquiries to be made in inactive address space without contention. This 'solicit successor' facility allows new nodes eventually to join the round robin. It is possible for a fault condition to develop which knocks out a node. The preceding node will be unable to pass on its token and after a number of abortive tries it will use the solicit successor function to identify the next active node. In this way the faulty node is dealt out of the system.

The token bus is undoubtably operationally more complex than previously discussed options but its deterministic action is valued in industrial environments where guaranteed worst case delivery times are vital. Probably the best known implementations are manufacturing automation protocol (MAP) and ARCNET

3.5.5 Polling

The underlying principle of polling is that a node may transmit only when it is asked to do so. The original local area network system of this type was synchronous data link control (SDLC). It was developed by IBM as a terminal cluster net emanating from terminal servers connected to a higher performance backbone network called system network architecture (SNA). Figure 3.13 shows the layout.

The importance of this system is not so much in its current use in local area networks but in the way that derivatives of it have been used as the layer 2 access protocol for connecting to public data networks. It will be useful to understand it when considering long distance interconnection of networks in Chapter 6, where a variant called link access protocol B (LAPB) plays an important part in carriers X-stream and integrated services digital network

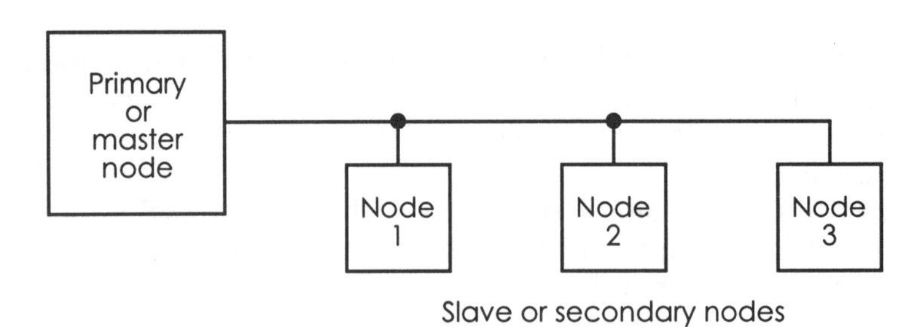

Figure 3.13 SDLC layout

(ISDN) options. Notwithstanding use as an access protocol, polling is a viable though not particularly efficient access method for local area networks.

It was initially stated that a node could not transmit unless it had been asked whether it wished to do so. The 'asking' is a function of a 'master' or 'primary' node. This master node asks each other, or slave, node in turn. By slight adjustment of the control field in the short enquiry frame the master can turn the enquiry into a 'selection'. This is necessary because the master acts as an intermediary in all frame transfers. To appreciate this action, consider the following scenario:

- Suppose node 3 has a message for node 7.
- Node 3 must wait until the master (node 1) sends it an enquiry frame.
- On receipt of this frame, node 3 sends a data frame back to the master with a destination address 7.
- Node 7 is not able to deal directly with any other node but the master and so the master 'selects' node 7 and relays the frame it has just received from node 3.

Efficiency is somewhat improved by allowing multiple frame transmissions and the use of goback *n* as a repeat strategy. The principles are exactly as described in Chapter 4, Figure 4.5. SDLC allows for three types of frame. Two are so called 'supervisory' frames, which include the polling and selection frames. Polling and selection frames are un-numbered. There are supervisory frames which are numbered and there are pure data frames. Acknowledgement does not use a special frame but 'piggy backs' on the polling response an expectation of the next frame number to be received. In this way, a node might imply that it does not have any data to send but if the master has one for the node it is expected to be number *n*. Such an idea provides for inherent acknowledgment, since saying that the next frame to be received is *n* inherently positively acknowledges frame $n - 1$. Figure 4.3 shows the format of an SDLC frame.

During periods of traffic inactivity a continuous polling cycle takes place between the master and the slave nodes. The master also controls the status of the other nodes and is able to switch them on or off line. Bit stuffing as described in Chapter 4, Section 4.1.2 is used.

The style of operation so far described is appropriate to the original multi-drop application. When an end-to-end only link is required, advantage may be taken of a variant specification called high level data link control (HDLC). Some of the improvements include the ability to request repeat of a specific bad frame, and the 'asynchronous response mode' which allows a slave to respond to other than a poll frame. This 'evens up' the status slightly, allowing a greater ability for the slave to initiate actions.

3.5.6 The token ring

The ring topology does demand a slightly different way of tapping into the highway. Figures 3.14(a) and (b) show the effect of the node/highway tap. It will be noticed that when nodes are 'listening' the ring highway is 'passed through' with only a single bit memory in series with the line. The purpose of this bit is to allow a passing bit in a data frame to be toggled or changed from 0 to 1 or *vice versa*. When a node is sending, the node circuitry opens the highway, as demonstrated by Figure 3.14(b). The single bit memory per node also acts equivalently to increase the number of bits that can be physically present on the highway. There is a good reason for this. The token in a token

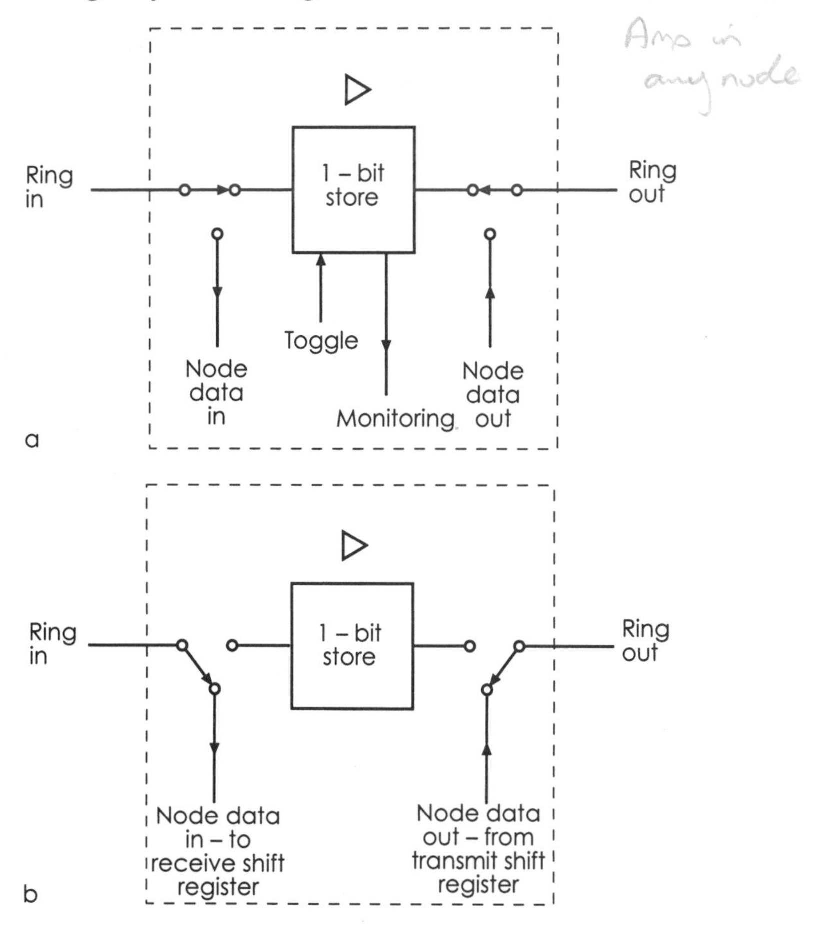

Figure 3.14 Token ring node (a) prior to and (b) during transmission

ring system is likely to be a short binary sequence and the total of the number of bits present in all of the single bit memories, plus the bits actively propagating along the medium of the highway, must equal the length of the token. If this is so, the whole token may be resident on the ring and continuously circulate. A short token frame is only a token frame whilst it conforms exactly to specification. If just one bit is changed it is no longer a token. This fact is used to advantage in the working of the ring which is now described.

If all nodes are idle the token circulates continuously. If a node wishes to transmit it will initially be listening and it will have assembled a data frame in its transmit shift register (see Figures 3.14(a) and (b)). In the listening mode it will be monitoring the passing data and will recognise the unique sub sequences in the token. It will be able to 'seize' the token by using the toggle feature of its single bit in line memory. The token is thereby removed and the next and subsequent nodes in line will not see it. Following the token removal, the sending node changes the configuration of its node/highway interface to that shown in Figure 3.14(b) and outputs the data frame from its transmit data register. Data frames on rings are often portrayed rather like toy trains on a circular track. This gives a poor impression of the action because the ring is likely to accommodate only a few bytes at most. A data frame may comprise thousands of bytes. The real picture is shown in Figure 3.15.

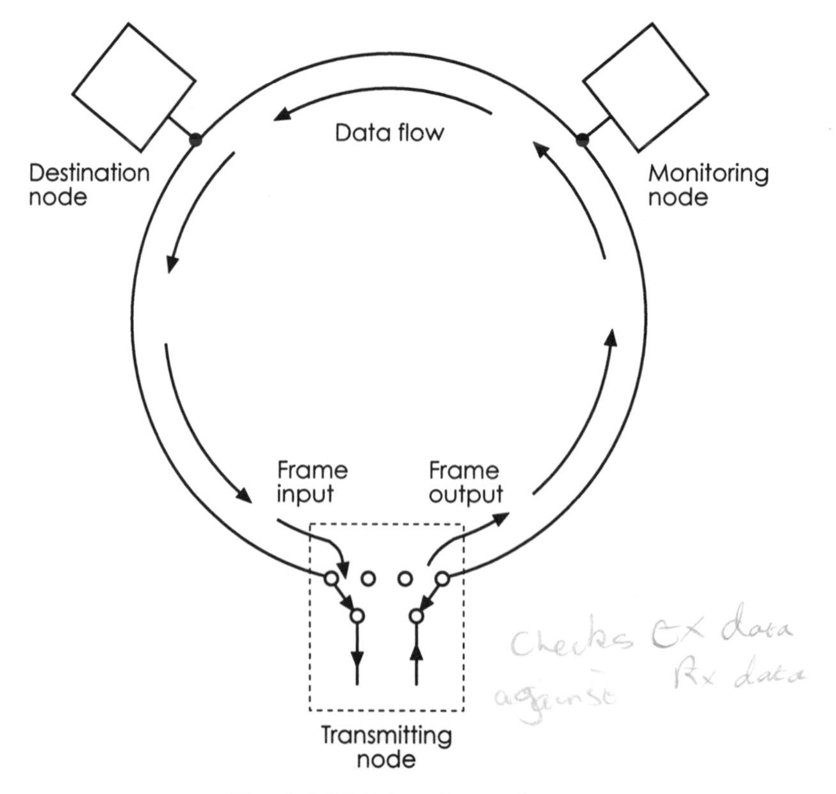

Figure 3.15 Token ring action

After a few bytes have been transmitted, the leading ones are back. This explains the design of the node/highway interface because the returning bytes must be removed from the ring as others are sent. Breaking the highway allows data to be clocked out of the transmit data register and also returning data may be clocked into the receive data register. The key point is that the sending node also automatically removes a data frame from the ring. Remember that only one node can send: the one that has seized the token, and all others are listening. As the frame propagates around the ring it passes through the series single bit memory of the destination node. The destination node is also monitoring the data that passes and will recognise its own address in the destination field of the frame. Having done so, it will read the frame into its receive data register as the frame passes. Near the tail of the frame, a single bit is reserved for acknowledgment and as this bit momentarily becomes resident in the series one bit memory, the receiving node will toggle it to indicate correct reception. When the frame arrives back in the sender's receive data register a quick check will show acknowledgment of receipt or not. If no acknowledgement is present the sending node may repeat the data frames a number of times. Of course, if no acknowledgement is then received the sender must desist. The protocol rules will prescribe what is then to be done.

Once a frame or group of frames has been sent and received the protocol will insist that the node possessing the token must relinquish it. The node regenerates a token and sends it onward along the highway where it may be seized by subsequent nodes. Because the protocol specifies the maximum number of frames that can be transferred for one token ownership event then, if the number of nodes on the highway is known, it is possible to determine a worst-case delivery delay for any data frame. A token ring can break down if the token is damaged or not regenerated properly, perhaps because of a fault. It is usual for one node to act as a network monitor. Since the worst-case time for a token to circulate is known or is calculable, the monitor node has an internal watchdog timer. If the time 'times out' then the monitor regenerates a new token. If it has consistently to regenerate tokens it can flag a fault condition to the network manager or maintenance staff.

This predictability and stability under even the heaviest load conditions together with the elegance and efficiency of acknowledgment have made the token ring a popular choice.

3.5.7 The contention ring

Imagine a ring constituted as described for the token ring above. In this case however, no token circulates. If a node wishes to transmit it may listen first and then switch its node/highway interface to transmit. If no other node is transmitting coincidentally, then the sending node receives its own frame back with acknowlegement just as with the token ring. Where things differ is when two or more nodes transmit approximately simultaneously. This may happen for the same reasons that were explained for CSMA/CD. A data collision is detected because the two (or more) sending nodes do not receive their 'own back'. This is illustrated by Figure 3.16.

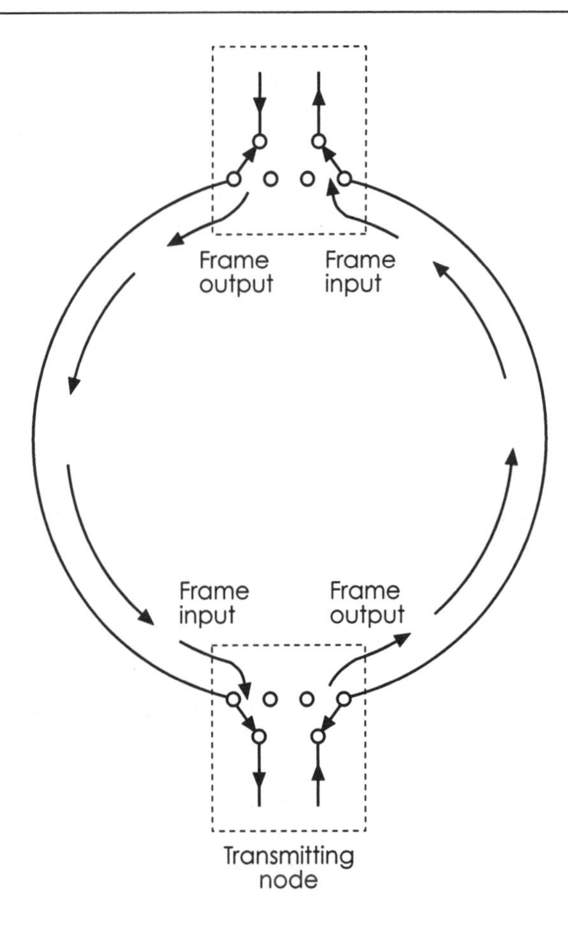

Figure 3.16 Contention ring collision action

3.6 Case study conclusions

The original specification for the network implies fairly heavy use but the environment is, or is similar to, an office. There is no imperative for fully deterministic action. The suggested staged implementation, starting in the headquarters building, provides an architecture that would seem to be well suited to Ethernet 10base5. The signalling speed of 10base5 is 10 Mb/s and this, together with the reduced and more even delays inherent to slotted contention systems promises to be adequate. The maximum size of a 10base5 highway is 500 m which is more than enough range for a single floor of the building. To wire the whole building, it will be necessary to interconnect the highway segments for each floor and provide a link between them which could be installed in the building lift shaft. Chapter 5 will explore the details of network interconnection and complete the validity of the suggested conclusions. Although the basic operational system has been chosen there are a number of practical issues of implementation.

3.7 Practical implementation: hubs and other things

Whilst a theoretical bus network may be used for conceptual models an actual installation has constraints and common sense adaptions of the basic idea. The decision for the case study was that an Ethernet style system was appropriate but for completeness the same practical considerations will be covered for a ring option.

3.7.1 Ethernet style pragmatism

Very often installations consist of clusters of client machines or terminals. Office environments are typical of this characteristic arrangement. Using a purely theoretical approach the network would be installed by ensuring that the bus passed close by each terminal and a separate transceiver tapping would be made for each. The costs would be significant and savings can be made by having a single connection to the bus, with the terminals connected to a terminal server as shown in Figure 3.17.

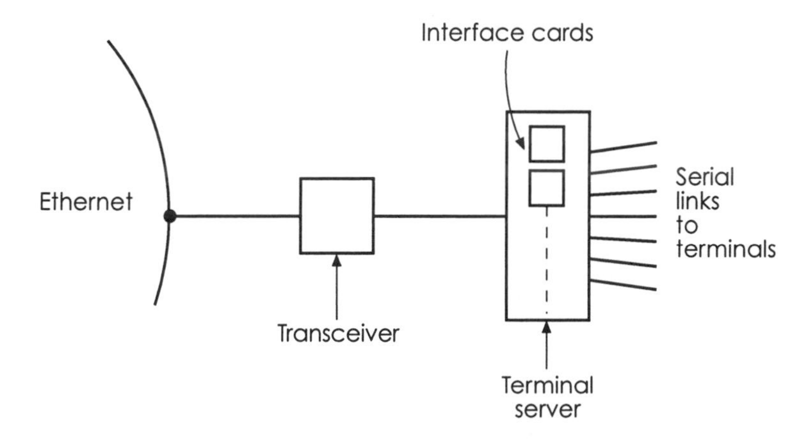

Figure 3.17 Terminal server

It should be noted that this option does not require each terminal to have an Ethernet interface. Ordinary serial synchronous or asynchronous links are adequate. The argument against this applies when some of the terminals are client computers or one of them is a file server. If either or both of these is the case, then a network action is needed between the terminals or computers. An ingenious solution is an 'Ethernet in a box'. Figure 3.18 illustrates the arrangement. Each computer or terminal requires an Ethernet interface and is connected to a 'local' Ethernet within a box. The box acts as a fan out box and could be used as a stand alone device for a small network or may be connected by a single transceiver (repeater) to a passing Ethernet highway. Large clusters such as one comprising 32 or 64 machines can be set up as cascaded arrangements as shown in Figure 3.19. It is possible to see how such fan out systems could be viewed as hubs.

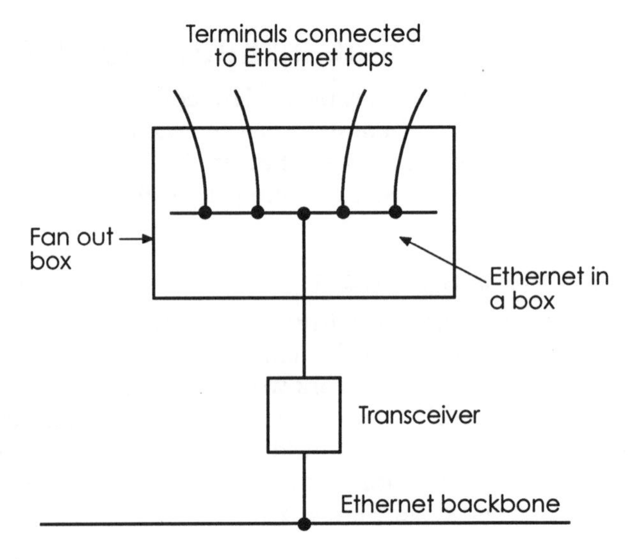

Figure 3.18 Ethernet in a box

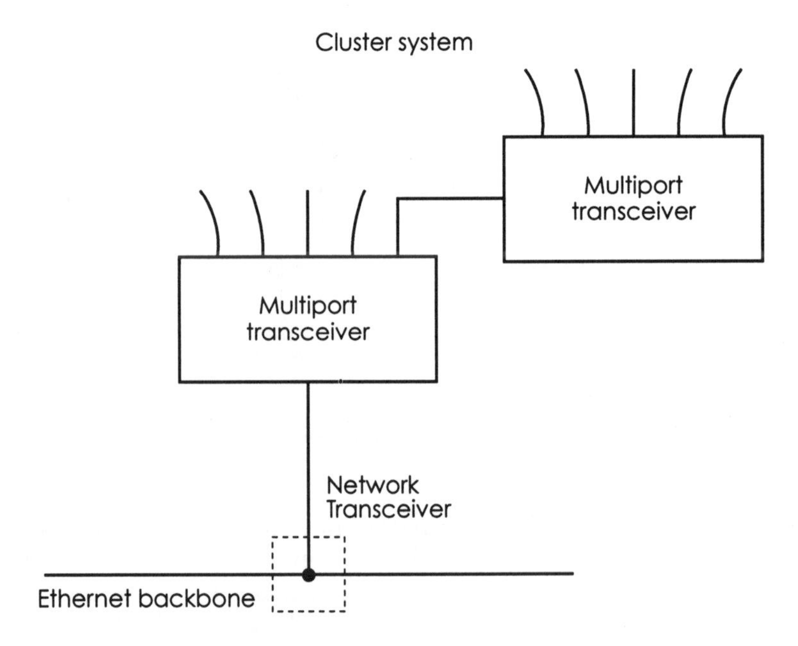

Figure 3.19 Large cluster system

3.7.2 Ring pragmatism

A similar kind of logic is present in the layout design of ring networks. Multistation access units (MAU) are used to structure the cabling and provide easy maintenance. An MAU is a section of 'ring in a box'. The box front panel has a number of sockets into which cables from each terminal or computer can be plugged. The ring enters one end of the box and exits the other. Figure 3.20 shows a single MAU. Physical centralising of the ring connections provides a superior system especially when faultfinding becomes necessary.

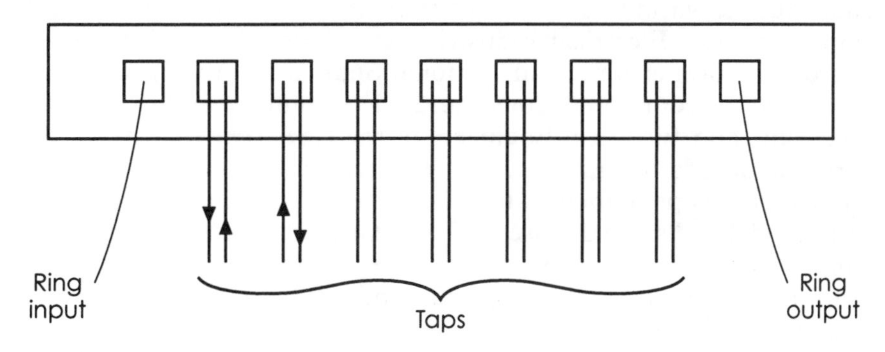

Figure 3.20 MAU

To understand the advantages consider the alternative or theoretical option where the ring cable passes close by each of the computers or terminals. If a fault developed, the testing might involve personnel crawling into awkward places and if a bypass was necessary it might prove very difficult to install. On the other hand the presence of a central MAU allows personnel to access a single place, make tests, and even bypass links by disconnecting them. This notion is so powerful that large hubs are created by racking together a number of MAUs as shown in Figure 3.21.

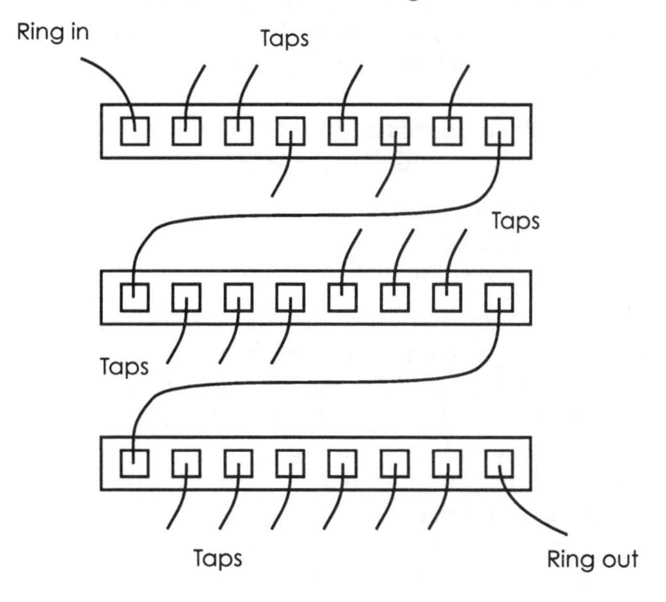

Figure 3.21 An MAU hub

3.8 IEEE standards

Prior to the recognition of the need for standards for interconnecting networks, controlling sessions and higher level data communications processing, it was necessary to define standards applicable to single local area networks. Throughout the world there are many standards authorities apart from ISO. Examples include: Electrical Industries Association (EIA); IEC and DIN standards in Europe; American National Standards (ANSI); CCITT, the international telecommunications standards committee and the Institution of Electronic and Electrical Engineers of the USA (IEEE). The IEEE is a powerful standards organisation and was early in establishing networking standards. The scope of their standards generally cover just layers 1 and 2 of the ISO model and since the use of IEEE standards has become universal it is important to consider the detail of the variants. As was explained in Chapter 1, the standards are called 802 standards because they were defined by committee 802 of the IEEE.

The IEEE saw fit to separate details of the data frame operation from the access method. The logical link control layer is defined in standard 802.2 and specifies just two types, type 1 and type 2. Type 1 is a connectionless method. This means that there does not have to be a previously established data link and little attention is given to numbering frames, flow control and error recovery. Type 2 does require a link to be established and numbers all supervisory and information frames to provide the services that type 1 lacks. Some unnumbered frames are used in type 2 to set up and clear down a link. IEEE 802.2 does specify frame formats and the protocol of frame exchanges. Inevitably it has its own jargon, calling a source node address, source service access point (SSAP) and a destination node address, destination service access point (DSAP). The basic frame structure is quite conventional. The frame begins with fields for source and destination addresses (SSAPs and DSAPs), a control field which is very like HDLC in nature, followed by an information field and a CRC.

The remaining IEEE standards define the relationship between the standards (802.1), access methods and data packet structures. They are as follows:

1. 802.3 – a standard describing a network which is CSMA/CD.
2. 802.4 – a token bus system.
3. 802.5 – a token ring system.

There is no inconsistency between ISO and IEEE standards and it is common to use IEEE options to implement the lower layers of a network which is designed on the basis of the overall ISO system.

Ethernet is an example of 802.3 whilst manufacturing automation protocol (MAP) is a classic example of building a full seven layer ISO system onto the lower layers implemented in an IEEE standard. MAP uses IEEE 802.4, a token bus, because of the deterministic nature of token operation. Determinism implies a guaranteed worst case delivery and response times for data. This is essential in machine control since probabalism is a problem when the network might carry machine emergency stop messages.

Questions

1 Compare and contrast ring and bus network topologies and illustrate that access method and topology are not linked. To satisfy this requirement, explain the action of:

(i) a token bus
(ii) a contention ring.

2 'In local area networking the access method to the data highway can be controlled by orderly rules or by contention.'

Comment on the truth of the above statement and expand your answer to explain the difference between deterministic and probabalistic access schemes.

3 Explain synchronous data link control, describing the background, facilities provided and protocol involved in data frame transfers.

4 Describe the consequences and subsequent recovery action when more than one station transmits simultaneously in a CSMA/CD network. Explain the role of the binary back off strategy employed after a data frame collision.

5 Speculate on the relative effects of long or short data elements being transmitted in a CSMA/CD system.

4

Error free channels

4.1 Data frame structures

4.1.1 Character oriented frames

The data blocks introduced in Chapter 1 and further justified in the context of synchronous signalling in Chapter 2 must now be examined in detail. The most natural structure would enclose or frame the message data by fields carrying the control information necessary for delivery and error detection. You will recall that character oriented blocks or frames make use of reserved ASCII characters. The ASCII allocation table is shown as Table 4.1: note that the non-printing characters below hexadecimal 20 include:

SYN – word synchronising character
SOH – start of header
STX – start of text
ETX – end of text.

The place of these characters and the purpose some serve as delimiters is seen in Figure 4.1, which illustrates a typical character oriented data frame, IBM Bisync. It should be understood that tables of association other than ASCII may well be used and in addition there is nothing sacrosanct about frame structures. Different designers will produce differing frame constructs. To illustrate the point, Figure 4.2 shows an alternative which is used by ARPANET.

The idea behind data frames is the provision of specific message entities that can be checked for errors and fully accounted for. This last point is

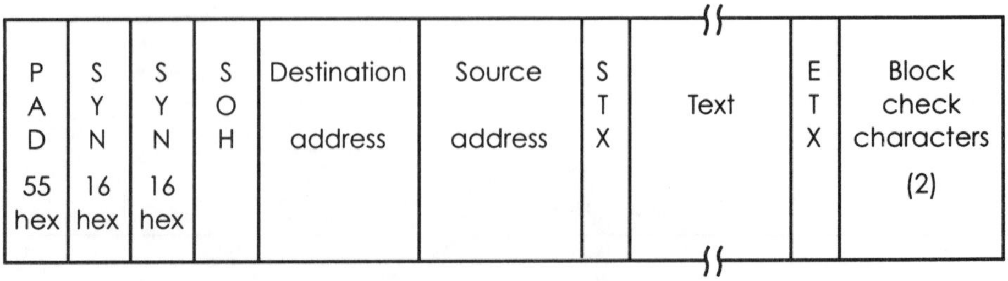

PAD 55 hex	SYN 16 hex	SYN 16 hex	SOH	Destination address	Source address	STX	Text	ETX	Block check characters (2)

Figure 4.1 Bisync dataframe

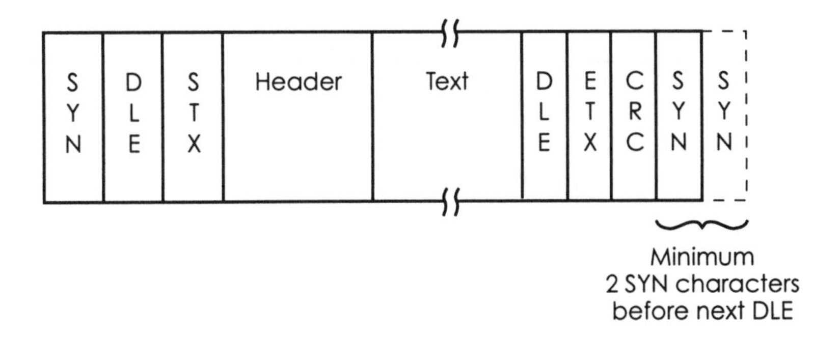

Figure 4.2 Arpanet dataframe

responsible for the fact that most data frames are numbered. The process of accounting includes the acknowledgment or otherwise of receipt. The rationale is evident from the function of ISO layer 2, which is to provide an error free channel. Error detection and correction, accounting and acknowledgement are all part of the achievement of this ideal. Implementation requires that network nodes behave according to a comprehensive and agreed set of rules and procedures for the exchange of data frames. These 'rules for dialogue' are collectively known as the layer 2 protocol which is discussed in Section 4.2.

4.1.2 Bit oriented frames

The definition of bit orientation is that each data frame may contain any number of bits. There is no constraint that only an integer number of character words may constitute a frame. No delimiting characters need be used but one unique binary pattern is required. The unique pattern is called the opening or closing flag because of its presence at the beginning and end of the data frame. If no special measures were taken in processing the data between the flags serious error situations might arise. This is because the arbitrary appearance of the flag pattern within the frame, as part of the message, would cause the receiving node to think that it had detected the closing flag. To stop this possibility, the transmitter undertakes 'bit stuffing', as shown in Figure 2.13. The flag pattern can no longer appear except as true opening and closing flags. The receiving node 'de-stuffs' the data between the flags in order to recover the original message.

There is need of fields within the data frame in both character and bit oriented systems but the difference is that the latter does not use delimiters between fields. A typical bit oriented frame from the IBM synchronous data link control (SDLC) system is shown in Figure 4.3. Typically, control fields are used for sequence numbers and acknowledgments. Utilisation of address fields differs with detailed design but might be expected to carry slave source and destination node addresses. The data field may contain arbitrary information in any number of bits, to a maximum defined. Maxima between 1008 and 128 kbits may be found in current products.

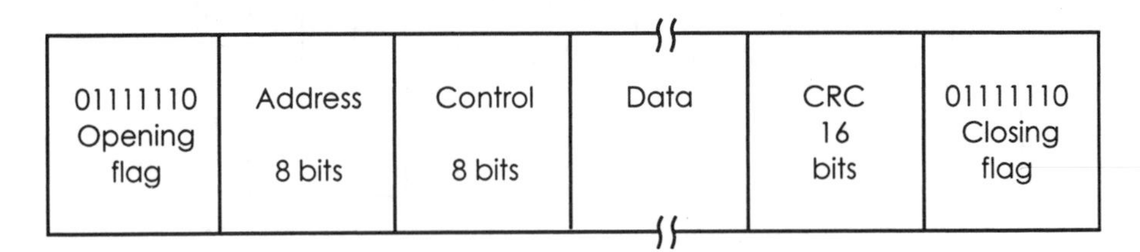

Figure 4.3 SDLC frame

4.2 Protocol issues

To understand the importance of protocol design decisions, consider a simple scenario. A number of data frames are to be passed between two nodes on the network. There is no sequence numbering of frames, flow control or error control. The lack of flow control is serious, since some processing time for receiving each frame is necessary and the transmitting node could issue frames faster than the receiver can deal with them. No error control means that corrupted data frames cannot be detected and the lack of sequence numbers could mean that whole frames could go missing or for as yet unexplained reasons get out of order and the receiver would be completely unaware of the problem.

4.2.1 Stop and wait protocols

Stop and wait protocols impose strong control of flow. After sending a data frame the sending node must stop and wait for a feedback frame from the receiver. The feedback frame need not carry any information in the simplest model, but since the feedback frame must be sent for further progress to be made, the frame could carry a positive or negative acknowledgment. The receipt of positive acknowledgment (ACK) by the sender allows the sending of the next frame. Negative acknowledgement (NACK) means that the sender must repeat the last frame. This protocol is not yet of sufficient quality to provide a fully transparent error free service. To understand why, consider what must happen if a data frame, or an acknowledgment frame is lost or corrupted by interference. One might suggest that after sending a frame and not receiving any feedback, a sending node should repeat the data frame. If the data frame had actually already been received and the acknowledgment frame had been the lost item, then the subsequent acknowledgment frame might well satisfy the sender. Unfortunately, the receiver will have received the same data twice without knowing. The output from the network will not be the same as the input and so the protocol will fail. Figure 4.4 illustrates the point. The receiver needs a mechanism for determining that what it sees is either new or a repeat transmission. The simplest solution is for the transmitting node to tag data frames with sequence numbers. By examining the numbers, the receiving node can differentiate as necessary. Having to acknowledge each frame can expose any processing latency in the system and it would be more satisfactory if acknowledgment could be given flexibly, accounting for a group of frames. Sending groups of frames without requiring individual acknowledgment is dubbed 'pipelining' and such a strategy requires the receiver to accept frame numbers within a given range or 'window'.

4.2.2 Sliding window protocols

There are two windows involved in sliding window systems. One is concerned with sending nodes and is effectively a list of frame numbers which the sender is permitted to send. The window size determines the action. If the window size is specified as 1, then only a straight sequence of frame sending can occur. If the window size is greater then the sender's list is a list of frames which have already been sent but for which acknowledgment has not been received and frames which have not been sent but may be. A sender's window will vary in size since a new frame may be added to the list at any time and a frame acknowledgment may arrive similarly. The window is like a queue with head and tail pointers. When a new frame is added the head pointer advances and when an acknowledgment for a frame arrives the tail pointer is advanced. A maximum window size will be specified so that once the limit is reached no more new frames may be sent until some acknowledgments are received. This gives an opportunity for the queue to be implemented as a rotating structure with, for example, allowed frame numbers being 0 to 7. When the head pointer reaches 7 the very next frame is allocated number 0, provided the tail pointer has moved on sufficiently. Figure 4.4 illustrates the idea.

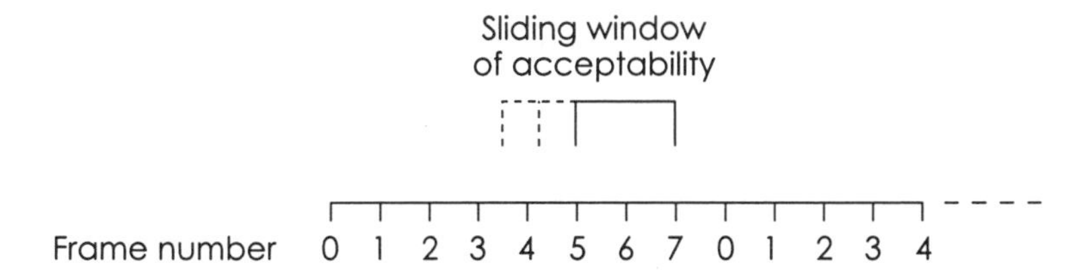

Figure 4.4 Queue for sliding window protocols

Receiving nodes also operate a sliding window, but in this case the window size is fixed and defines the frame numbers that the receiver is permitted to accept. Any frame outside the acceptable range is usually discarded. A frame bearing the lowest acceptable frame number is received and the window moves on. As an example, imagine a receiver window of 3 which currently defines as acceptable frame numbers 2, 3 or 4. When frame number 2 is received an acknowledgment is generated and the window moves on to allow frame numbers 3, 4 or 5. This window too may operate as a rotating queue.

Although the system permits frames to be received out of strict order it must be emphasised that the peripherals connected to the network nodes must never be aware of this fact. In other words, the node software must only present the peripheral with frame contents in precisely the order that it was given to the transmitting node by a host computer or peripheral.

What are the advantages of increasing the protocol complexity in this way? In non-windowed protocols frame 2 cannot be sent until frame 1 has been acknowledged. Windowing allows pipelining or the grouping up of frames for transmission. If the window size for a given system were 8, then a stream of frames up to that maximum may be sent and a single acknowledgment given. Pipelining frames over a network which is occasionally vulnerable to noise or data corruption will mean that sooner or later a frame in the middle of a group may be corrupted. In this case general acknowledgment cannot be given and a repeat strategy is needed to arrange for the retransmission of the corrupted frame or indeed frames.

4.2.3 Repeat strategies

There are two popular repeat strategies, go back *n* and selective repeat. In go back *n* the receiver only acknowledges the frames prior to the data corruption event. Figure 4.5 shows the action. A group of three frames is transmitted. During frame 2 a burst of interference destroys the validity of the frame data. The receiver will detect this and issue an acknowledgment for frame 1 only. On receipt of the acknowledgment the sending node knows that a data corruption has occurred in frame 2, simply because if 2 had been received correctly the acknowledgement would have been for that frame. When frame 2 is acknowledged it is implied that frame 1 is also correctly received. In response to a situation like this, the sending node repeats all frames after the explicitly acknowledged one. This means that the sender 'goes back *n*' frames. It could be argued that this method wastes network bandwidth by repeating some frames that were correctly received in the first instance. In the example, frame 2 is corrupted but frame 3 may well have not been. Nevertheless both frames 2 and 3 will be repeated in the go back *n* strategy. The argument would be valid when a network is subject to high error rates, for instance, a manufacturing environment.

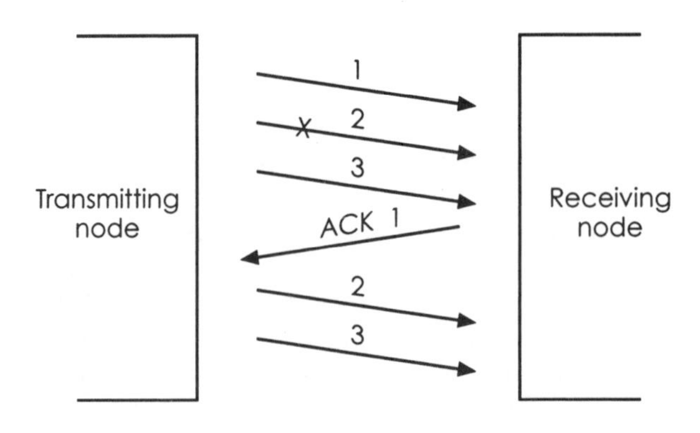

Figure 4.5 Go back n repeat strategy

Selective repeat requires the receiving node to store incoming groups of frames, and in the event of data corruption, indicate the erroneous frame. The sending node then only repeats the frame in question. The receiver replaces the faulty frame in its store before yielding up the full data to the attached host or peripheral. In deciding the most appropriate choice all performance factors of the network and node costs must be considered. The decision is based on a trade-off between node memory and the bandwidth of the data highway.

4.3 Error detection

4.3.1 Parity checking

In asynchronous serial communications it has been traditional to include a token gesture to the detection of errors induced into transmitted words. The system used is called 'parity' and a parity bit is included between start and stop bits (see Figure 2.2). Before transmitting a particular word the number of 1s in it is determined. If even parity is being used, the parity bit is set to 0 or 1 in order to make the total number of 1s even. For example, the ASCII code for character 'A' is 41 hexadecimal, 01000001. Because two 1s are present the parity bit is set to 0. Hexadecimal 58 is the code for 'X', 01011000. This time the parity bit must be set to 1 in order to create four 1s.

Odd parity is equally permitted but obviously both ends of a link must agree on the system in use. The real problem with this is that, although it is simple, it does not detect compensatory errors. For example, consider ASCII 'A' once more. 01000001 has a parity bit 0. So does 01001000 or 10000001. The system will only detect errors in a single bit or three bits or five bits and so on.

4.3.2 The use of a block check character

It has been established that networks are best served by synchronous systems which carry data in the form of blocks or frames. With a character oriented system it seems reasonable to include a parity bit within each signalled character as before. A significant opportunity exists to exploit the existence of the data frame to improve on basic parity checking. The best way to appreciate the concept of block parity is to imagine the characters making up the block or frame laid out one underneath another. Figure 4.6 demonstrates the principle. The frame is considered to consist of six characters but it will be noted that seven are shown. The last character is generated by creating its bits from the calculation of vertical parity. This means that the first bit of each character word contributes to the associated columnar parity bit. The last character word is called the block check character and it includes a bit which is a vertical parity of parity bits. The individual parity bits are a horizontal or row parity whilst the block check character provides vertical or columnar parity. In this way, the system is based on a two-dimensional parity which vastly increases its effectiveness when compared with the simple word parity alone. Careful experimenting with Figure 4.6 would show that the only errors that are not detected are those that occur as two bit errors in the same column, in the same block. Error detection is based on the principle that the block check character is calculated at the sending end and appended to the block or frame. The receiving end performs the same calculation and should end up with the same block check character. Comparison identifies whether or not the frame has suffered corruption.

Word	7 bit code							Even parity bit	
1	0	1	0	1	1	1	0	0	
2	1	0	0	0	1	0	0	0	
3	0	0	1	1	1	0	1	0	
4	1	0	1	0	0	1	0	1	
5	1	1	1	1	0	1	1	0	
6	0	0	0	1	1	1	0	1	
	1	0	1	0	0	0	0	0	Block check character

Figure 4.6 Block check character

4.3.3 Cyclic redundancy check

Cyclic redundancy check provides high performance error detection. It is based on the creation of a near unique 'signature' which characterises a given string of 0s and 1s such as might be found in a bit oriented data frame. A cyclic redundancy check (CRC) generator is based on the creation of a pseudo-random binary sequence. A shift register with feedback is used to operate on the data within a frame. The register involves a number of single bit delays, each of which may be defined by using an operator D :

$$X(t) = DX(t-1)$$

Multiplying by D is equivalent to delay by 1 bit and the CRC generator shown in Figure 4.7 may be described by :

$$D^3X(t) + DX(t) + X(t) \text{ or simply } X^3 + X + 1$$

The polynomial represents the system. As data bits from the frame are clocked into the system the register fills up. When the register is full, clocking a new bit in will result in the loss of a bit from the end. As data bits clock through the register, their states may affect the next input because some register bits have a feedback path which allows them to reach the XOR conditional input gate. The content of the register thus reflects previous content or history and the echo of history reverberates in the system. Once all the data bits of a data frame have been clocked into the system the remaining content of the register is a near unique number which characterises the bit pattern of the frame. The transmitting node applies the algorithm and tags the CRC signature onto the frame within the error check field. On receipt, the receiving node applies the same algorithm and should arrive at the same signature. It compares its own calculation with that of the sending node and if the same result is found,then the data frame has been received correctly. It will have been noticed that the term 'near unique' has been used to describe the signature. This is because there is a probability that the same signature could result from different bit streams. It can easily be demonstrated that the possibility is very small with practical signature and frame sizes. Although the principle has been explained using a 4 bit register the reality is that a typical real world network protocol might use a 16 bit word. The probability of failure of the system to detect an error is :

$$\text{Prob(fail)} = 2^{(m-n)}-1/2^m-1$$

Where m is the frame data length and n is the size of the register. So, for a 16 bit register and a 256 bit data frame the probability of failure to detect an error is only 1.526×10^{-5}. Another way of thinking of the process is to remember that in modulo-2 arithmetic no carries exist for additions or borrows for subtraction and hence addition and subtraction are simply exclusive OR. The process of long division involves shifting and subtracting and so the system in Figure 4.7 may clearly be thought of as a long division. The polynomial describing the system is called the generator polynomial and it is divided into the data frame bit stream treated as a number. A more realistic generator polynomial is :

$$X^{15} + X^{12} + X^5 + 1$$

These polynomials can be expressed in binary (the one above is: 1001000000100001) and so the requirement is simply to divide one binary

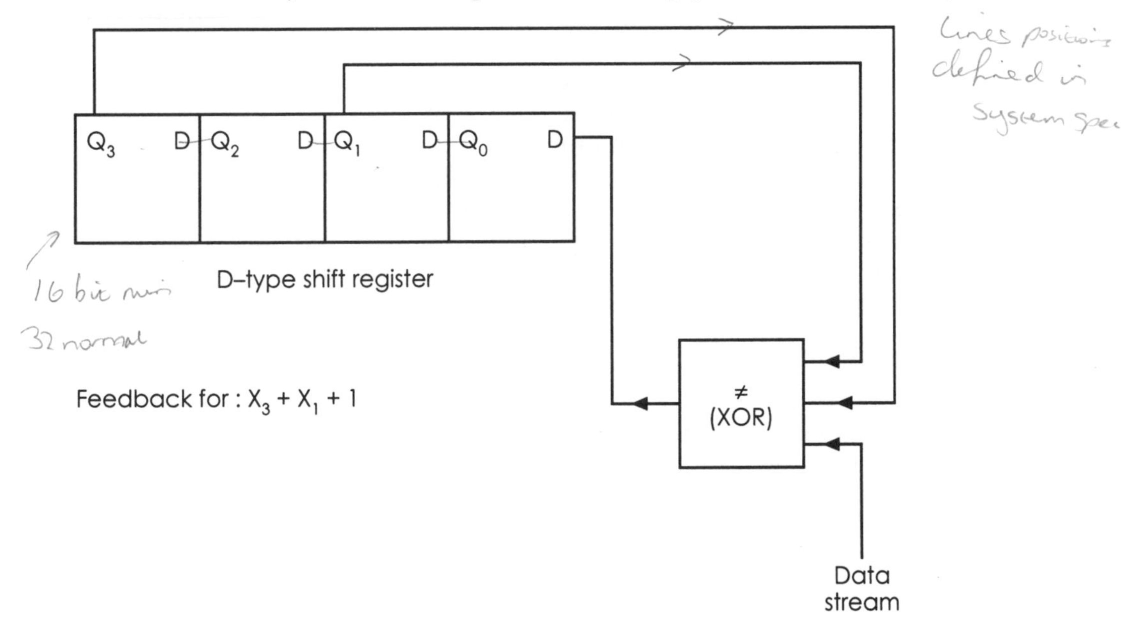

Figure 4.7 CRC generator

number by another. When this has been done, there will be a remainder. It is the remainder that remains in the register of Figure 4.7 and is the signature required. A 16 bit remainder or signature will ensure that it is possible to detect:

- All single and double errors
- All errors with an odd number of bits
- All burst errors of length 16 bits or less
- 99.998% of 18 bit and longer burst errors.

A burst error comprises a number of corrupted sequential bits in a data stream. They are very common in data communications. The performance of CRC is excellent in dealing with these problems.

4.4 Error correcting codes

Everything that has gone before in Section 4.3 assumes that it is only necessary to detect the existence of errors and that repetition of the erroneous frame will be taken care of by the protocol in use. This is a perfectly reasonable assumption for most networking and communication systems. There are a few applications where extreme range may result in unacceptable delay when data has to be repeated. In these cases, codes are used that are sufficiently robust that even when a relatively large number of bits has been corrupted it is still possible to determine what the message actually was with certainty. Inevitably such codes involve introducing redundancy, which provides an interesting counterpoint to the aims of data compression covered in Chapter 9. It is sometimes possible to use compression first, followed by error correcting coding so that the worst effects of the extra redundancy are mollified.

The use of error control coding in a given case would need to be the subject of careful evaluation in terms of 'cost-benefit' analysis. Calculating the 'costs' of coding is a well understood process.
To correct t errors for a given artificial added redundancy, m ($t<2^{(m-1)}$), then n, the overall block size, is2^m-1 and $n-k = < mt$.
k is the size of the information field of the block. As examples consider a code where n, k = 31,11. With an overall block size of 31 and an information field of 11 there are 20 parity digits (mt).

$2^m = 32$ therefore $m = 5$ and since $mt = 20$, $t = 4$.

A 31,11 code can thus correct four errors. Consider a code 63,45. In this case $2^m = 64$ and thus $m = 6$. With $mt = 18$, in integer terms $t = 3$. A 63,45 code corrects three errors.

4.4.1 Essential background

It will be useful to describe terms and concepts generally applicable to codes suitable when data is broken up into blocks such as data frames in networks. Not surprisingly this class of codes is called block codes. The terms and concepts applicable to block coding will now be examined.

4.4.2 Galois fields

In many instances applications of algebraic thinking are required to be implemented by digital computer. In this context it is important that, whatever is done, the rules of ordinary arithmetic must apply in order that algebraic techniques can be used.

A Galois field defines the constraints in terms of number symbols and operations within which this is true. For binary work the alphabet 0 and 1 are defined along with the basic arithmetic operations of modulo 2 addition and multiplication. This set or field of elements is usually denoted GF(2). Galois fields must have 2^n symbols,where n is a positive integer.

4.4.3 Cyclic codes

Cyclic codes are those where if an n-tuple

$$v = v_0, v_1, v_2 \ldots v_{n-1}$$

is a code vector, then the n-tuple

$$v^i = v_{n-i}, v_{n-i+1}, \ldots v_0 \ldots v_{n-i-1}$$

obtained by end around shift (one place right) is also a code vector.

4.4.4 Polynomial representations

It is possible to treat each code vector as having a one to one correspondence to a polynomial of degree $n - 1$ or less.

With polynomial representation it is possible to develop some important properties for a cyclic code which make implementation of encoding easier.

Included in these properties are:

1. In an n,k cyclic code there exists only one polynomial $g(X)$ of degree $n - k$.
   $$g(X) = 1 + g_1X + g_2X^2 + \ldots g_{n-k-1}X^{n-k-1} + X^{n-k}$$
2. Every code polynomial $v(X)$ is a multiple of $g(X)$ and every polynomial of degree $n - 1$ or less which is a multiple of $g(X)$ must be a code polynomial.

4.4.5 The generator matrix

To generate each code word a generator matrix G could be used instead of a look up table. The code word is arrived at by matrix multiplication e.g.

$$[v] = [m][G]$$

where v is an $n \times 1$ column matrix and m is a $1 \times k$ matrix, thus G is an $n \times k$ matrix. This is best illustrated by example of a 'systematic' code. A systematic code is one where the codeword has two concatenated parts. The first part is the message word and the second part is a group of redundant digits which are a parity function of the message bits. Assume a message word 101. The redundant bits, say three, might be defined as follows :

bit 1	even parity bit for 1st and 3rd message bits
bit 2	even parity bit for 1st and 2nd message bits
bit 3	even parity bit for 2nd and 3rd message bits

This yields a redundant portion - 011 and therefore a complete code word of 101011

The code contains 3 message bits within a total of 6 bits and would be described as a 6,3 code.

Consider an encoder segmenting messages into 3 bit blocks, it then transforms each block into a code vector of 6 digits.

Message	Code
000	000000
001	001101
010	010011
011	011110
100	100110
101	101011
110	110101
111	111000

Because $k = 3$ there are $2^3 = 8$ possible distinct messages. Each is transformed into a unique codeword. The list of eight 6-bit words is a subspace of the total vector space of all 6-tuples. It is possible to find a set of linearly independent 6-tuples such that each 6-tuple is a linear combination of others. The whole code can then be described by a matrix and a codeword can be found by a linear combination of the rows of this 'Generator' matrix.

Using the 6,3 code example above a generator matrix for it is:

$$G = \begin{matrix} v_1 & 100110 \\ v_2 & 010011 \\ v_3 & 001101 \end{matrix}$$

Element v_1 corresponds to the message word 100, v_2 corresponds to message word 010 and v_3 corresponds to message word 001. All other codewords are found by modulo-2 addition of these.

Thus if the message is 101 then the codeword is :

$$(1 \times v_1) + (0 \times v_2) + (1 \times v_3)$$

$$= 1 \times (100110) + 0 \times (010011) + 1 \times (001101) = 101011 \quad \text{(modulo-2)}$$

Clearly the first three digits of the code word are the message bits, and the last $n - k$ bits are linear functions of the message bits. These $n - k$ bits are the parity check bits and the code is systematic. Instead of storing the full eight codeword options at the encoder only the 3×1 generator matrix is necessary in order to generate the full set. The code is specified by the generator matrix.

4.4.6 The parity check matrix

The transpose of the generator matrix is the parity check matrix; e.g. for the generator matrix above the parity check matrix H is :

101100
110010
011001

This is obtained by the transposition being specified as below:

v_1= 100110 bits 1...5
v_2= 010011
v_3= 001101

is read bottom to top, right to left, yielding:

$$H = \begin{matrix} 101100 \\ 110010 \\ 011001 \end{matrix}$$

Where H is the parity check matrix.

The usefulness of the parity check matrix is to allow the derivation of the syndrome.

4.4.7 The syndrome

Consider the case of a message word 111 from the generator matrix; the code word is 111000. If the transpose of H (i.e. H^t) is derived by converting rows into columns it is :

110
011
101
100
010
001

If the code word 111000 is denoted as u then a matrix multiplication can be carried out as follows :

$s = u.H^t = 110 + 011 + 101 = 000$ (modulo 2)

s is called the 'syndrome' and will be 0 for all genuine codewords, but not if corruption has taken place. For example consider the case where the codeword 100110 has been corrupted into 101110. The syndrome will be :

$s = 110 + 101 + 100 + 010 = 101$ and is non-zero.

Having detected the errors it is necessary to identify them in order to correct. One way of doing this is to use the standard array method which allocates a one to one correspondence between non-zero syndromes and the correctable set of error patterns.

Each formulation of this type of code has a different error correcting capability. The 6,3 example is only capable of correcting single bit errors. Using the notion that a correct codeword can be changed into a corrupted codeword by the modulo-2 addition of an error vector, then the error vectors that can be corrected by the 6,3 code above are :

000001
000010
000100
001000
010000
100000

This is the 'correctable set' of error vectors, otherwise known as the coset leaders. Six non-zero syndromes could be associated with these coset leaders and simple interpretation will allow error correction for these error vectors.

If a corruption occurs for which a coset leader is not responsible then an incorrect decode will result. Systems should be devised so that the coset leaders are the most likely errors to occur.

The details above are general and apply to any n,k linear block systematic code. The term 'distance' is used to define the number of components by which two code words differ. The 'minimum distance ' for a code is the smallest distance between all possible pairs of codewords. The notion of minimum distance is important in that it determines the error correcting capability of a linear code. Greater than 1-bit error correction is possible but an n,k of more than 6,3 is needed. For instance a 15,5 code corrects up to 3 error bits.

4.4.8 Bose-Chaudery-Hocquenghem codes

This code is generated by means of a generator polynomial. This is derived by letting 'a' be a primitive element of Galois field (2^m). A primitive element is any element whose powers generate all the non-zero elements of GF(2^m). The least common multiple of the minimum polynomials of a^i.

The form of the generator polynomial for GF(2^4)
is as follows :

$$g(X) = 1 + X^4 + X^6 + X^7 + X^8$$

yielding a 15,7 cyclic code able to error correct 2 bits. Other possibilities can be created by using more or fewer of the minimum polynomials. Thus using the three minimum polynomials generates a 15,5 code that corrects 3 bits, whereas using the single minimum polynomial is single error correcting.

The decoding of BCH codes assumes that the code vector has been added to by a noise vector. The first step of the decoding process is to calculate the syndrome which is a vector with $2t$ components, using the received vector. The set of equations is solved for each of the vector elements and then a technique is applied which finds an error location polynomial. Finally, the roots of this last polynomial are found which identifies the error positions within the received vector.

4.5 Convolutional codes

The codes described in Section 4.4 are all block codes. This means that they operate on a short block of information. The size of the block was dependent on the error correcting capability required. There is an alternative approach which seeks to generate code on a continuous basis. The idea is that data to be encoded is input one bit at a time, but continuously. Parity check digits are created 'on the fly' as a function of the current and n past digits. The total of the digits considered in the continuous data stream is called the 'constraint length'. One way of establishing a given constraint length is to clock input data into a shift register. The number of bits in the shift register is the constraint length. A set of parity generators monitors different combinations of the shift register bits and the data bit is followed by $n - 2$ parity bits. Each time a new bit is entered into the system, it passes through but has appended to it a number of parity bits. Monitoring for parity is a relatively simple function. Assuming even parity the function is provided by a binary half adder for each parity bit. With a constraint length of four the system required is as shown in Figure 4.8. The half adding of bit 1 of the register simply means

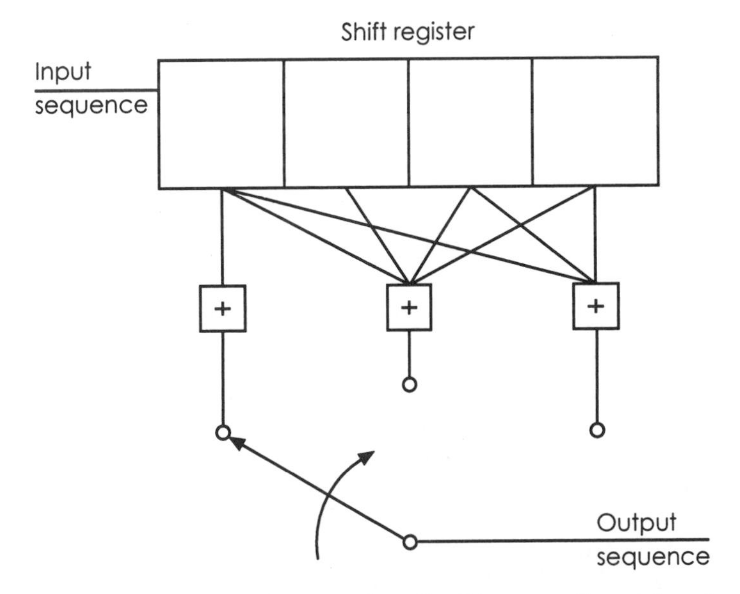

Figure 4.8 Convolutional coder after Wozencroft and Jacobs

that the first output is the original data bit. Consider a five bit input stream, 01101. For the system as shown, the output for the first bit input and for one cycle of the output switch will be: 000, assuming a cleared register at the start. When the next input bit is clocked into the shift register, the output will be 111. Continuing produces the sequence 101,001,111,001,011,011,000. In this way 01101 is coded as 000111101001111001011011000. A further assumption that has been made is that the register is 'flushed' with 0s once the word to be encoded is fully entered. The flushing gives rise to a residual code tail which in this case is 001011011000.

Coding is unique to a particular size of register and the connections to the adders but for each design it is possible to create a code tree. For the system in question the codetree is given in Figure 4.9.

For decoding, data is treated in sections proportional to the constraint length and the receiver may 'look up' the intended message by reference to the code tree. Progress is made by a process of highest likelihood, for example, in Figure 4.9, the first branch split is 000 or 111. Given 101 as the first three digits of the encoded message, the greatest likelihood is that the

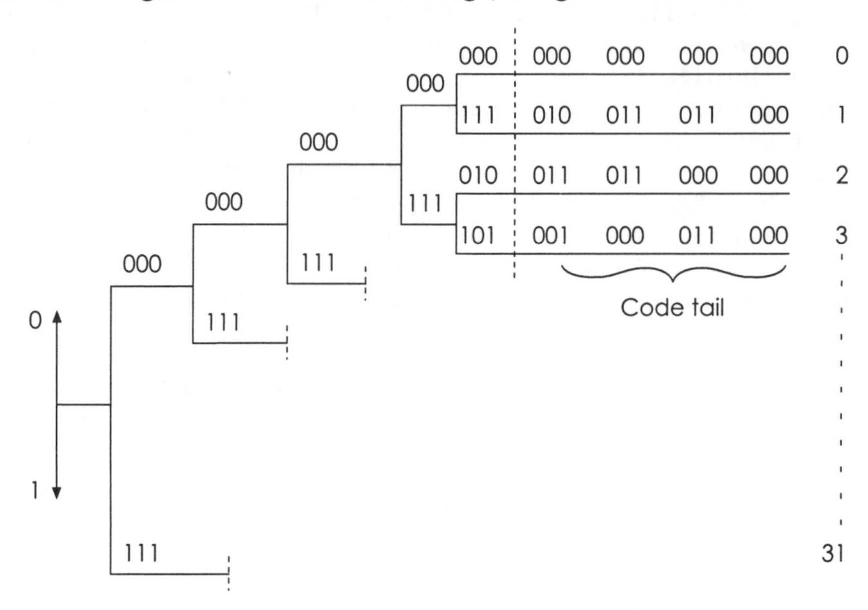

Figure 4.9 Convolutional coder part codetree

code should have been 111 but has been corrupted. A kind of cumulative likelihood function is computed and decisions made. If more and more differences arise as the tree is progressed then, once past a specified threshold, it is possible to back up and try a different pathway through the tree. Decoding is thus a probabalistic business, but nevertheless works extremely well and is used in much telecommunications equipment intended to work over noisy channels. The decoding method described is effective and is called the Fano algorithm. An alternative and faster approach called the Viterbi algorithm recognises symmetries and repetitive parts of the tree and redraws the tree as a lattice. Additionally, all the pathways are searched simultaneously in the Viterbi method.

4.6 Conclusions for the case study

In Chapter 2 we concluded that a bit oriented system was a wise option considering the open ended possibilities for the nature of data. This has been confirmed as appropriate in this chapter and it would only be necessary to agree the exact form of the data frames. As far as protocols are concerned, the robustness and resilience of the sliding window option makes it a natural choice. The go back *n* repeat strategy will be chosen because of its simplicity and because it is not expected that the proposed network application will fully utilise available bandwidth. There appear to be no distance or delay justifications for the added complexity of error correcting codes.

For error detection the clear leader is the cyclic redundancy check. No real problems exist in obtaining repeats over the network as used and specified.

Questions

1 Discuss the relative merits and application areas of: parity, block checksum; and cyclic redundancy check error detection methods.

2 Explain the principle of operation of cyclic redundancy check (CRC) and show by means of example how:

(i) The error detection bits are generated
(ii) A received frame is checked for errors

Assume the generator polynomial $x^4 + x^2 + 1$ and an eight bit message 11100110.

3 By drawing frame sequence diagrams illustrate the effect of the send window flow control limit being reached in a link level protocol. Assume a send window of 2 and a go back *N* error control procedure.

4 In the context of error control strategies, analyse the relative power and performance of:

(i) Character parity in asynchronous transmission
(ii) Block check methods for data frames
(iii) Cyclic redundancy checks for data frames.

5 Differentiate between block systematic codes and convolutional systematic codes in error correcting systems.

5

Interconnecting networks

5.1 Overview

Connecting networks together may be achieved in a number of ways and at a number of levels. The networks that are to be joined may operate autonomously or they may not. They may be running the same or similar protocols or they may not. The distances involved may be local, national, international or global. Later chapters will tackle the range of general techniques available for longer range working whilst this chapter concentrates on the principles and issues of interconnection relevant to networks operating locally or the range from a few tens of kilometres to national distances. Local internetworking is appropriate to the case study since the policy was planned as a staged development. After the initial installation, the network manager will be adding new networking features and there will be a need to integrate new services which may have varying degrees of match with the existing system.

The main method of interconnecting networks is to use a relaying device which interfaces to conditions found on each side. A useful basis for discussing the functionality of this relay device are the layers of the ISO OSI model and this will be done.

5.2 Repeaters

If two networks are to be interconnected and they are running exactly the same protocols, frame formats, access methods and signalling strategies, then the problem is at its simplest. As an example, consider the decisions already taken in the fulfilment of the case study specification: a baseband system; synchronous signalling; Manchester binary modulation; slotted contention bus (Ethernet). The choice of Ethernet 10base5 (10 Mb/s, 500 m network length, 100 nodes maximum), satisfactorily allows the networking of one floor of the building in question. Four floors are involved and it is sensible to interconnect each floor. Since precisely the same network specification applies to each floor network then one of those 'simplest of problems' exists. The maximum range on one Ethernet highway often leads to the kind of problem here and because interconnection is so often needed, the individual highways are referred to as network 'segments'. With Ethernet the maximum distance between two communicating nodes on separate segments must not exceed 1500 m.

Segments are joined by a relaying device called a 'repeater'. Repeaters are only appropriate when networks that operate under exactly similar conditions are being joined, as in this case. Implementation in the case study calls for segments or highways that are a little in excess of 200 m long. The excess allows for part of one end of each segment to provide a short vertical run in the lift shaft of the building. Figure 5.1 assumes about 3 m and suggests that the repeaters reside in the lift shaft itself. This is perfectly reasonable since repeaters are not connected to any terminals or host computers and are completely automatic in action. Since they are intended to be 'out of mind' in operation, they might as well be 'out of sight'.

The vertical runs account only for a few tens of metres and thus the maximum range between nodes should be under 500 m where the nodes are at the far end of the top and bottom floors.

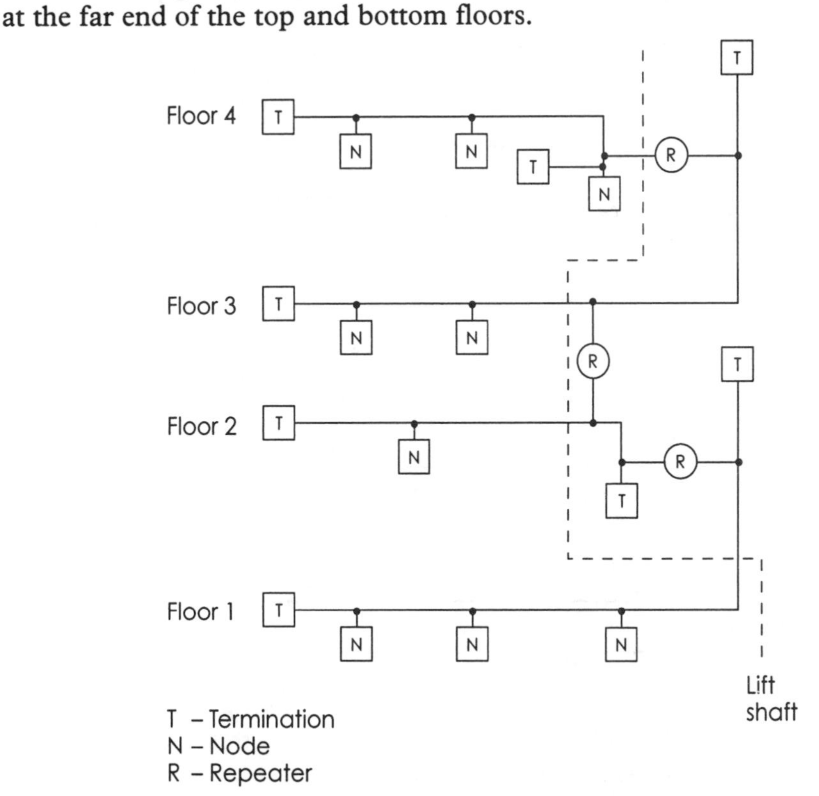

Figure 5.1 Interconnecting bus segments

5.2.1 Repeaters for contention bus systems

A repeater operates at layer 1 of the ISO model. It is there simply to repeat and boost the signals present on the segments the repeater joins. It is thus a two way device and it must be made clear that the use of repeaters means that the separate segments cannot be thought of as autonomous. A repeater does not have any significant memory and 'passes through' signals with no 'listening first'. There is no receipt of a data frame and then a separate contention action on the next segment. The repeater does not have this capability. It operates at the bit level. The result is that the introduction of repeaters makes the access method operate across the extended network as a whole.

5.2.2 Repeaters for other systems

Many network options, such as ring topologies, cannot use repeaters in the same way as contention systems. Indeed, it is not necessary. Each node in a ring has an amplifying or repeating function and because of this a ring can have a much greater overall length. Figure 5.2 illustrates a ring equivalent of the four floor case study described above. Note that the concept of segments is inappropriate.

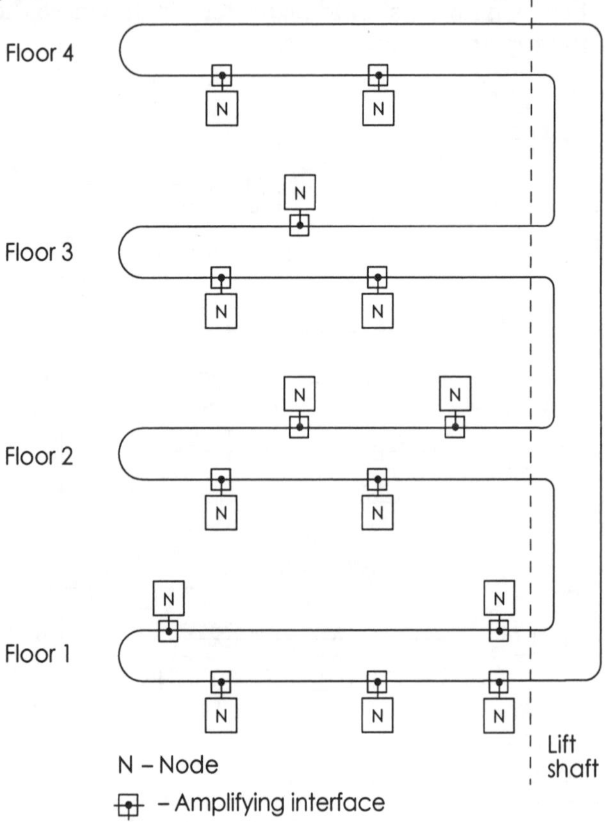

Figure 5.2 Ring installation

5.3 Bridges

Another form of relay device is the bridge. In this case, the device does act in a 'store and forward' way. A data frame from one network is received by the bridge and is then relaunched on the other network using the protocol of that network. The best way to imagine the way a bridge works is to think of it as two back to back nodes as shown in Figure 5.3. Bridges operate at ISO layer 2, the data link layer. This means that a received data frame is processed to remove the message carried in the data field and then inserted into a new data frame for onward transmission on the new network. Point 'x' in Figure 5.3 identifies where the raw message could be examined.

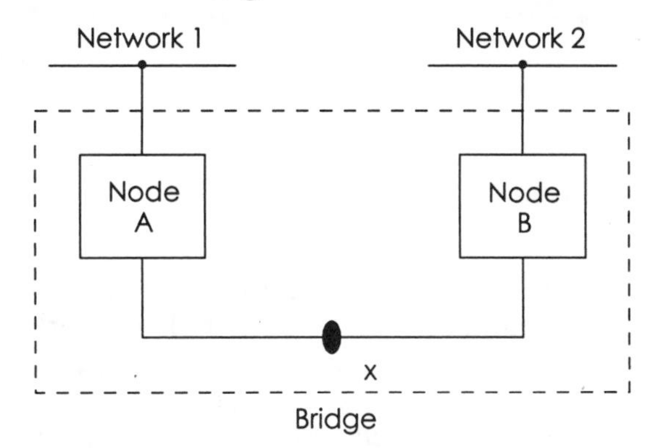

Figure 5.3 Bridge

When does a bridge relay a data frame? The bridge may be programmed to recognise all node numbers as being local to one side or the other. If a frame appears on one side but does not have, in its destination address field, a node number valid on that side then the bridge does its work. In other words it isolates the two networks and only passes non-local traffic.

This method of operation opens up the possibility that the two interconnected networks could be operating under different access methods, topologies, and protocols. Because a non-local frame is stripped down to a raw message which is then reinserted into a new frame it is also possible to use a bridge to overcome the restrictions on repeaters in the Ethernet system. In that way it is possible to exceed the 1500 m limit. A bridge is entirely appropriate for joining say, a token ring to an Ethernet or a 10base5 network to a 10base2 network. For some interconnections it is wise to provide the bridge with significant first in, first out buffer memory. This is necessary because the standard data rate on a token ring is 4 Mb/s, whilst for Ethernet it is 10 Mb/s. Figures 5.4(a) and (b) show some possibilities for bridge usage.

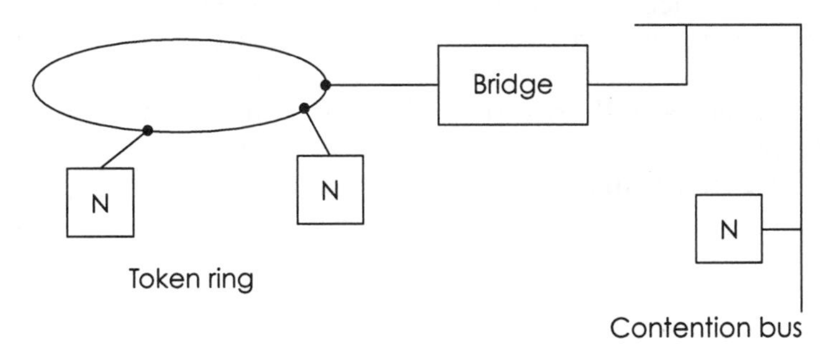

Figure 5.4(a) Bridging between dissimilar networks

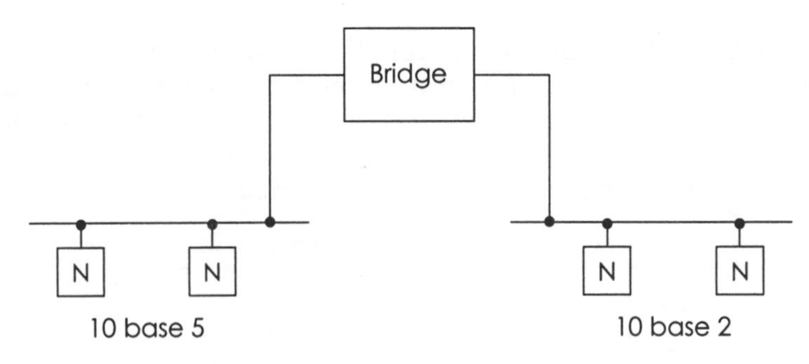

Figure 5.4(b) Bridging between generically similar networks

5.4 Routers

Yet another category of relay device is the router. With the repeater, the action of the device made the segments part of one whole network. Bridges joined possibly dissimilar networks but it was clear that bridges identified frames that were to be relayed because they did not possess 'local' addresses as destinations. This implies that networks joined by a bridge must have common administration and an agreed node addressing policy. What happens when it becomes necessary to join two similar networks that are under completely separate administration? The answer is that data frame data fields must carry not only pure data but also an address. In this way, when the frame is stripped away enough information is available to enable the route to the destination to be determined.

What is being suggested is that the content of the data field has a structure, with a header containing address and control information. This structure is called a data packet. Some people use the terms 'frame' and 'packet' interchangeably but this is incorrect. A packet is always fully contained within a frame, as was illustrated by Figure 1.4, during the discussion of the ISO 7 layer model in Chapter 1. Layer 3, the network layer, has the responsibility for network routing and addressing and it is at this level that the router operates.

Section 5.7 explores the main options for implementing routing policies.

5.4.1 Internet

Ideally, every network node in the world would have a unique address. Those nodes on purely internal networks that do not communicate with others arguably do not need a unique address. Internetworked systems would however clearly benefit. One very dominant standard for the format of data packets is the Internet packet (IP). Unique addresses for this standard are provided by a central authority in the USA. Network managers apply for and receive 32 bit addresses which may then be allocated to nodes on their networks. The addressing scheme is shown in Figure 5.5.

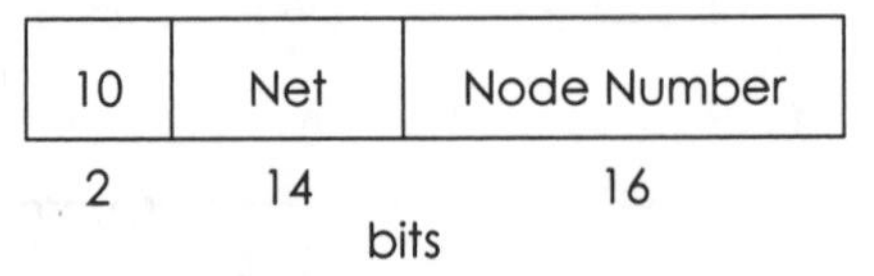

Figure 5.5 Internet addressing

The scheme allows for just over 4294 million nodes. Addresses are documented in a special notation. Each byte of the 32 bits is expressed as a decimal number and is separated by a full stop or dot. Examples of Internet addresses might be 174.23.77.102 or 234.3.45.99. Figure 5.6 illustrates the form of the Internet protocol packet header. The Internet operates by sending these packets as datagrams.

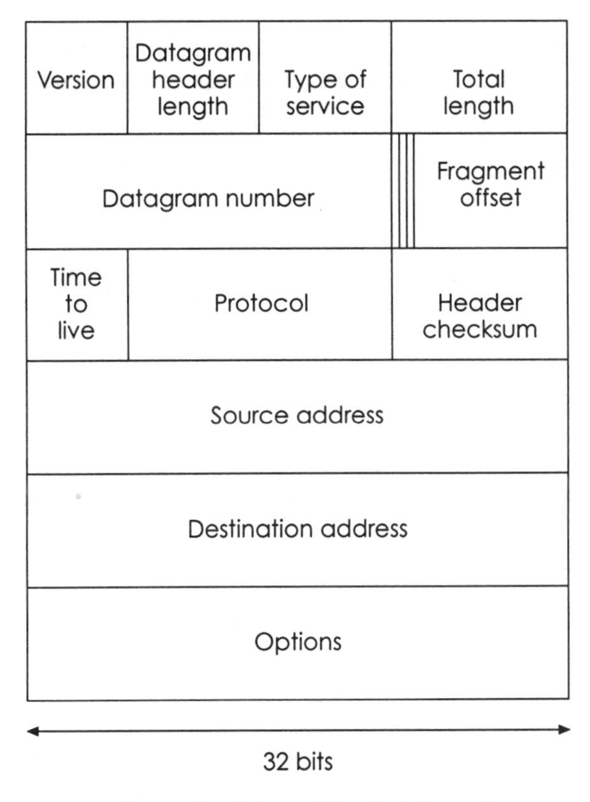

Figure 5.6 Internet Packet format

A few of the various fields are defined as follows.

Version

The Internet protocol is still developing. As a result it is possible for slight variations to occur in format. The version number allows easy identification and the possibility of packets being modified en route to suit the requirements of a certain link.

Datagram header length

The options field at the end of the packet header may consist of up to 32 words. It is a variable length field, which means that the header also has a variable length. One way to handle this is to state, up front, what the length is in a particular case.

Type of service

Internetworking can be be provided with services having different qualities. The need for reliable, accurate delivery might be applicable to some uses, while others might demand speedy delivery as the important criterion and not be concerned with the detailed accuracy.

Total length

This is overall total length of the combined header and message fields in the packet in bytes. The 16 bit size of this field sets the maximum total length at 65536 bytes.

Datagram number

During transit, packets may arrive out of order, especially when store and forward units are used to tackle congestion. The datagram number helps the receiving node 're- assemble' full messages from fragments received in different packets. Numbering is also an obvious necessity for acknowledgement.

Options

The options field may be used to carry many possible indicators such as: time stamping; security-encryption data; and error messages. It also provides a built in vehicle for adding and experimenting with features not defined in the original standard.

Time to live

This is an eight bit number which is decremented for every second that passes. An IP datagram may only live for a maximum of 255 seconds. When the counter reaches zero, the next network router reached will delete the packet. The time provided should be more than ample for a datagram delivery and the deletion feature automatically removes relics and undeliverable dross from the system.

5.4.2 X.25

Whereas the Internet provides a datagram service, an alternative packet level system is offered by CCITT X.25. The CCITT, a consultative committee concerned with international telecommunications, is a standards authority set up for the benefit of telecommunications public carriers such as BT, Mercury, AT&T, France Telecom and so on. The standards prescribed by the CCITT are used by the public carriers as the basis of their own wide area networks. X.25 is the access standard for connecting to the public data network and as such will be returned to in Chapter 6. Its relevence here is that it defines a data packet format that could be used by any network designer wishing to interconnect distant networks. X.25 has been quite popular as a private internetworking standard because of the consistency offered when private links and links via public carriers all use the same system.

X.25 is an example of a 'virtual call' packet protocol. Virtual call systems may exist in one of three possible states at any given time. As was discussed in Chapter 1 the analogy is that of a telephone call. The three states common to X.25 and a telephone call are: the call set up phase; the conversational phase; and the call clearing phase.

A node that wishes to communicate with another on a different network must 'call' the destination. When contact is established, data may be exchanged until there is nothing more to say, and then contact is broken. To achieve this, X.25 prescribes three packet formats: the call request format; the control packet format; and the data packet format. These are shown in Figures 5.7(a), (b) and (c). The call request packet is much larger than the control type and it may be seen that source and destination addresses (calling and called addresses) appear only in the request packet. When the request

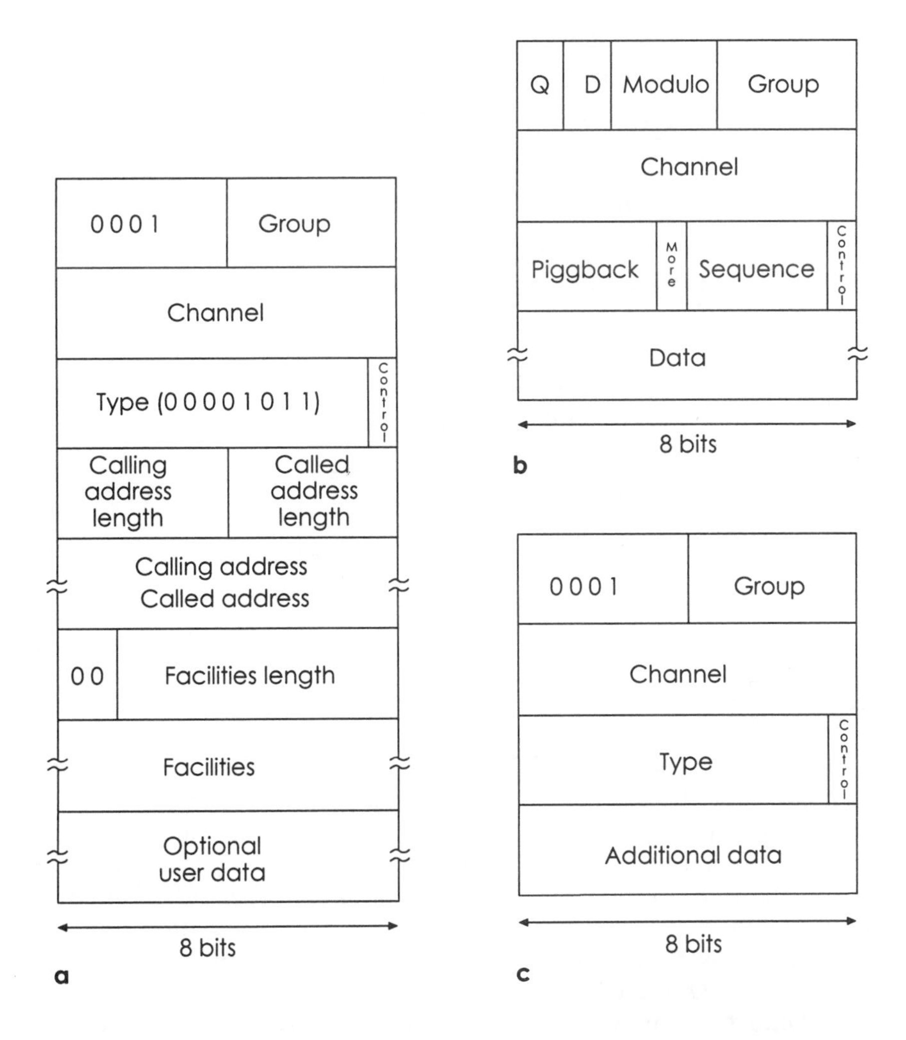

Figure 5.7(a) X.25 Call request, **(b)** Data Packet and **(c)** Control Packet

packet is generated an arbitrary 'channel number' is assigned. All subsequent packets refer only to this channel number. The assumption is that the router will create an association between addresses and channel number and thus it is unnecessary to keep supplying the full addresses after they have once been indicated. The router simply needs to know that a particular data packet is part of a particular 'call'. The packet format of X.25 is far shorter and simpler than that of the Internet packet, which because it is concerned with a datagram service, must always carry full delivery addresses. Figure 5.8 tabulates and lists the possible control packet types. A large variety is necessary to provide foolproof and orderly connection and disconnection of the calls.

Type	Code
Call request	00001011
Call accepted	00001111
Clear request	00010011
Clear confirm	00010111
Interrupt	00100011
Interrupt confirm	00100111
Receive ready	PPP00001
Receive not ready	PPP00101
Reject	PPP01001
Reset request	00011011
Reset confirm	00011111
Restart request	11111011
Restart confirm	11111111
Diagnostic	11110001

Figure 5.8 X.25 type fields

5.5 Gateways

Operating at the highest levels of the ISO 7 layer model, gateways are relay boxes intended for the interconnection of completely dissimilar networks. The protocols on either side need have nothing in common since the gateway accepts data under one protocol, completely removes all trace and then recodes the messages in the form and protocol of the other side. The lowest element of commonality between the two sides is likely to be the application program. The gateway can connect IBM token rings to DECnet, IBM SNA (System Network Architecture) and to MAP (Manufacturing Automation Protocol). Almost any network can be connected to any other, provided that the gateway is properly specified.

5.6 The fibre distributed data interface (FDDI)

The interconnection of networks up to 100 km apart is the aim of FDDI. The medium is optical fibre specified as 625/125, which means that the core diameter of the fibre is 62.5 micrometres and the cladding is 12.5 micrometres. The use of optical fibre gives a huge bandwidth enabling signalling to be carried out at up to 100 Mb per second. The role of such a system is to provide a very high speed interconnection or backbone network to which bridges and routers may be connected. FDDI operates as a token ring system, but with up to 500 nodes and up to 2 km between those nodes. The maximum frame size is 4500 bytes. The 100 km maximum range is hardly likely to be a fully privately owned highway and parts of the ring may well be implemented by renting optical fibre channels from public carriers. Of course some users will have much smaller rings and be using FDDI not for its range but for its speed of transmission. One curious feature of FDDI is the 'double counter rotating ring' architecture. The primary and secondary rings can be useful in overcoming faults in the highway. A faulty section can be removed provided that nodes are connected to both in parallel. The provision of two rings means that one may be used as a backup or they may be used as individual highways. The nodes for FDDI are classified as either 'Type A' or 'Type B'. Type A nodes are indeed connected to both rings whilst Type B are connected to the primary ring only. Another unusual specification relates to the signalling at the physical level. One of the design objectives of FDDI is to use the fibre bandwidth to transfer as much data as possible. If certain forms of binary modulation are used, such as Manchester encoding, then the two states per bit characteristic could be seen as a waste of potential performance. To minimise this loss, a 4 bit/5 bit code is used in which a 4 bit pattern is translated into a 5 bit pattern. Each of the 16 possible 4 bit patterns is linked with one of the 32 patterns available in 5 bits. The selected 5 bit patterns are all sufficiently assymetrical to provide synchronous clocking detection for the receivers. No long runs of 0s or 1s are permitted in the chosen codes. This scheme means that, to carry 100 Mb of actual data, sufficient bandwidth for 120 Mb per second is needed. By contrast, if Manchester encoding were used, 200 Mb per second of bandwidth would be essential. Typical usage of FDDI is shown by Figure 5.9.

In Figure 5.9, the 'concentrators' are convenient centralised wiring points which specifically allow Type B nodes to interconnect with the secondary ring. A concentrator is a specialised Type A node. An FDDI ring performs two kinds of token passing: the unrestricted token and the restricted token. The former may be seized and used by any node but the latter is only available to nodes that have been allocated 'priority'. Different nodes are allocated a proportion of the total available time. The allocation is dependent on the traffic expected to be offered by given nodes. This is implemented by a variable called 'rotation time'. In reality this feature is tied in with the handling of priority on the ring. There are many possible priority strategies that could be used but with FDDI the notion is that nodes likely to offer little traffic are given least rotation time and are allowed to seize tokens before others. The logic of this strategy is that least rotation time nodes are going to present little and not very often. Giving them a high priority avoids such nodes 'waiting' for large amounts of data to be shifted by heavy traffic nodes. Letting them seize the token causes only very little delay before they release the token again. Error detection is through of a 32 bit cyclic redundancy check (CRC). FDDI is a system designed to provide functionality up to the equivalent of ISO layer 2, and thus packet level operations are the responsibility of the networks and devices inserting and recovering data to and from the ring.

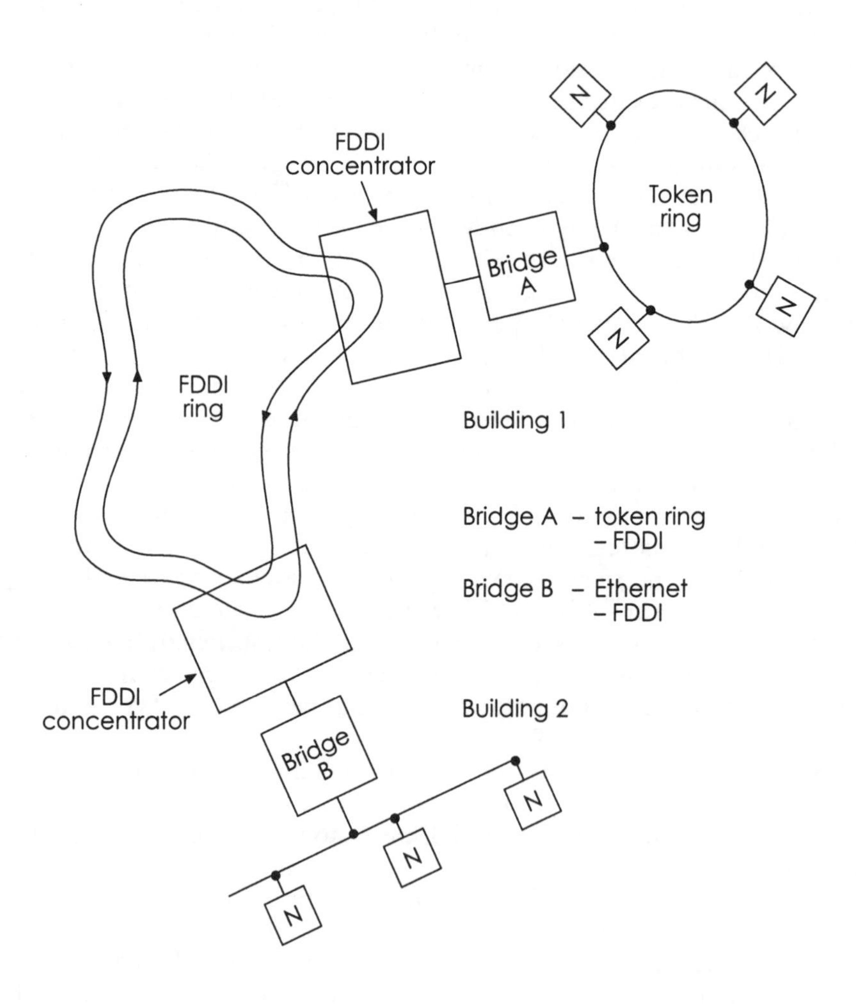

Figure 5.9 Typical FDDI usage

5.7 Asynchronous transfer mode – ATM

The very nature of much data traffic is that it may be described as 'bursty'. With the increase in multimedia computing, the bandwidth required for interconnecting sites will increase sharply. Whilst FDDI offers 100 Mb per second, predictions are that six times this will be necessary. The problem in providing for this is the burstiness referred to previously. For such high bandwidths it is tempting to supply simply a dedicated circuit for each channel, but the disadvantage of the circuit switching strategy is that bandwidth is wasted. Circuit switching is best when traffic is offered in a smooth predictable way. When traffic is generated sporadically and as packets, the idle times waste resources that other channels might benefit from. The proposed solution is to offer 'bandwidth on demand'. Data sources are expected to produce data bursts, effectively at any time, and the system would cope by virtue of its huge bandwidth and policing together with a 'data loss' fallback strategy.

5.7.1 ATM topology and operation

The topology of an ATM system is based upon the concepts of point to point networks. High speed data highways are connected through 'cross-over points' or switches. This structure has architectural implications, particularly for switch design. The random traffic capability inevitably means that congestion could occur and must be catered for. Traditional congestion control advocates the use of first in first out buffers and the architectural options for ATM switches provide various combinations of buffers, multiplexers and busses or crosspoints. Figures in the group labelled 5.10 rehearse the main possibilities, and Figure 5.11 illustrates a simple multiplexer.

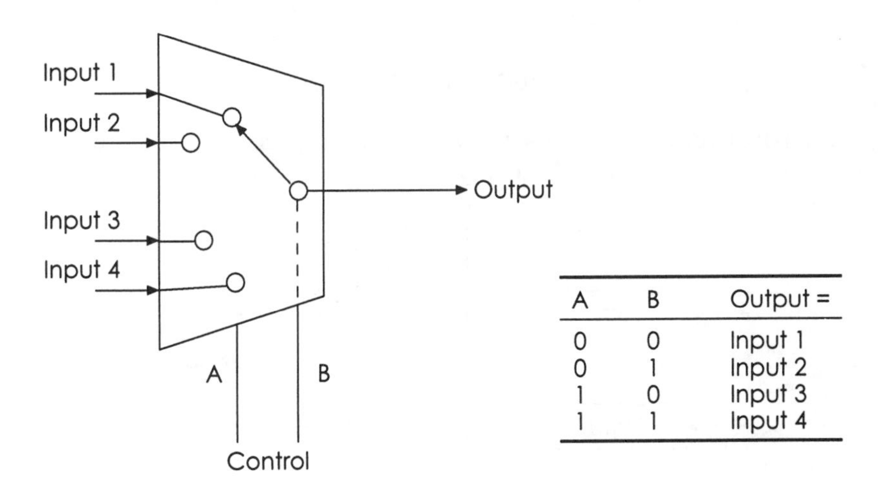

A	B	Output =
0	0	Input 1
0	1	Input 2
1	0	Input 3
1	1	Input 4

Figure 5.11 A simple multiplexer

In Figure 5.10(a), the output of the multiplexer is concentrated onto a parallel bus which operates in much the same way as a bus within a computer. The output buffers are addressed as ports, and incoming channels may be directed to a particular output port and subsequently a channel buffer. The bus is acting as a distributor. This option offers the advantage of simple control but is vulnerable to problems if and when the bus bandwidth is exceeded. The bus is a potential bottleneck.

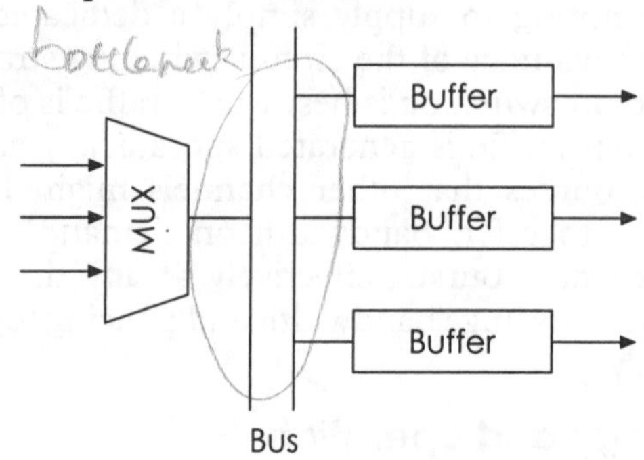

Figure 5.10(a) ATM output buffer

The shared buffer of Figure 5.10(b) offers a relatively small amount of buffer capacity but the control of memory is a complicated function. Traditional switching technology makes extensive use of arrays of crosspoints and

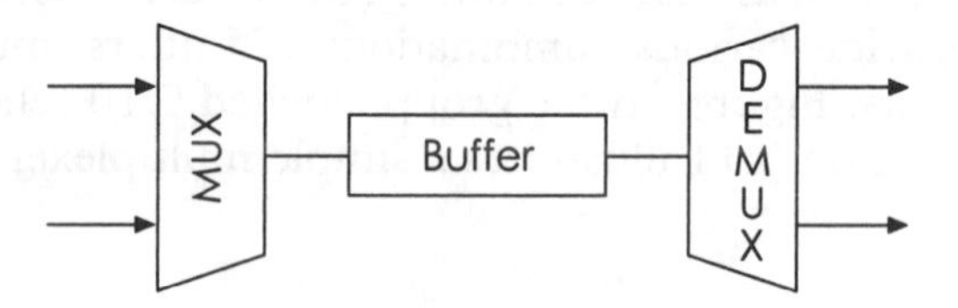

Figure 5.10(b) ATM Shared buffer

Figure 5.10(c) represents such a system. Closing contacts at the crossing points of row and column allow any input to be connected to any output.

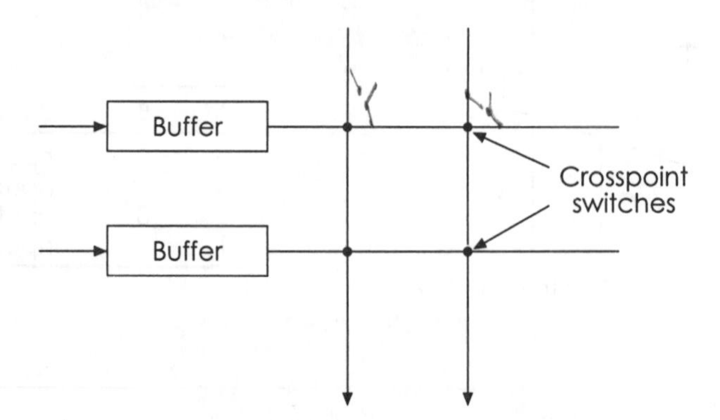

Figure 5.10(c) ATM Input buffer

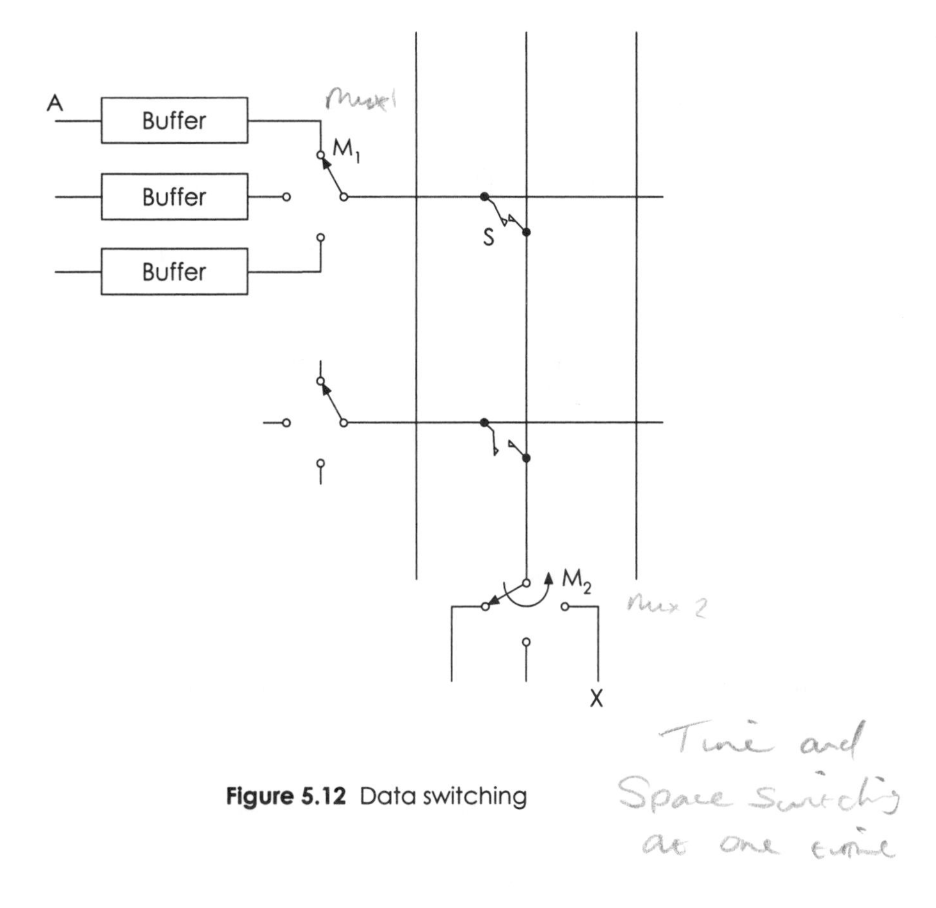

Figure 5.12 Data switching

Control of this kind of system is very complicated for packet switching because it involves switching in time and spatial domains. To understand this, consider Figure 5.12. To connect channel A to channel X, multiplexer switch M1 must wait until multiplexer switch M2 is correctly positioned. The input buffer allows for the delay and only when M1 selects A and M2 selects X should the spatial domain switch S connect the two together and allow the channel A buffer to empty onto the correct output pathway. Whilst this single example seems relatively easy the problems are thrown into sharp focus imagine the switching patterns when all input channels are loaded and if you require crossconnections to any combination of output channels.

The problems are reduced by the crosspoint buffer option shown in Figure 5.10(d). The distributed buffers provide a large buffering capacity and the control is relatively simple. Switches using output buffers tend to give the best delay and throughput performance. Switches using any form of shared buffer make best use of memory. Input buffered systems can be embarrassed if there is a long delay in making the correct switch connection and ATM does have a discard policy. Clearly, discarding must be kept within agreed limits in provided an acceptable quality of service. Typical expected acceptable loss rates are thought to range between 1 in 10^6 and 1 in 10^9. The choice of switch

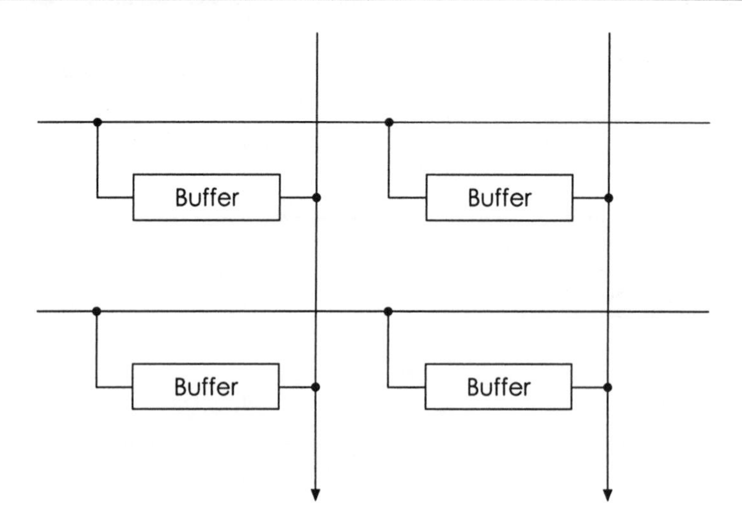

Figure 5.10(d) ATM Crosspoint buffer

design is a trade off between complexity, reliability, and modularity. ATM switches may consist of multiple stages such as the 'banyan' shown by Figure 5.13.

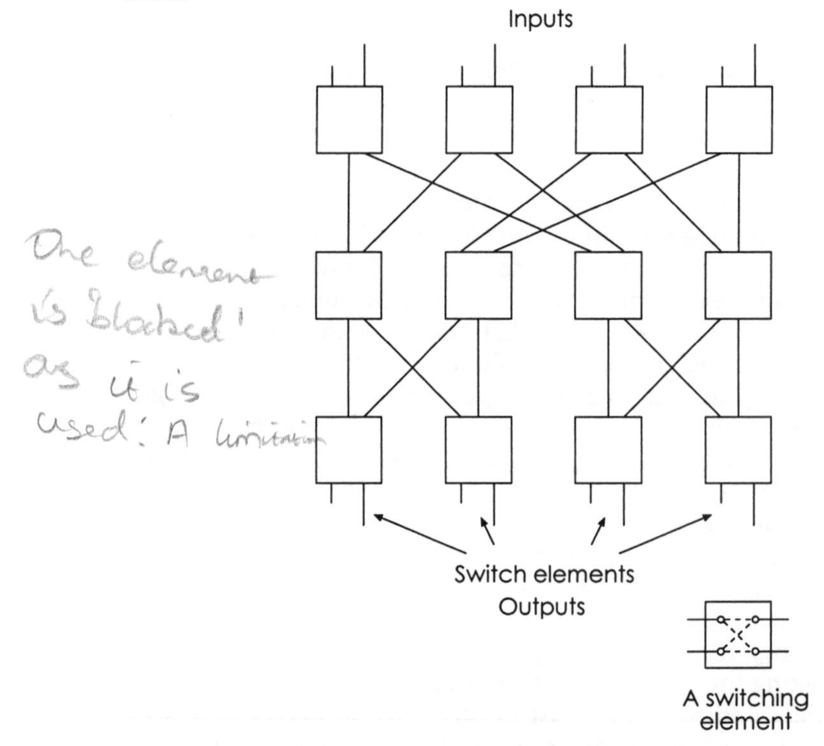

Figure 5.13 Banyan network

5.7.2 ATM Packets

The ATM terminology for a packet is 'cell' but for consistency it will not be used. The ATM packet has a fixed size of 53 bytes or octets. The header is 5 bytes and the remainder is a data field. The structure is as given in Figure 5.14.

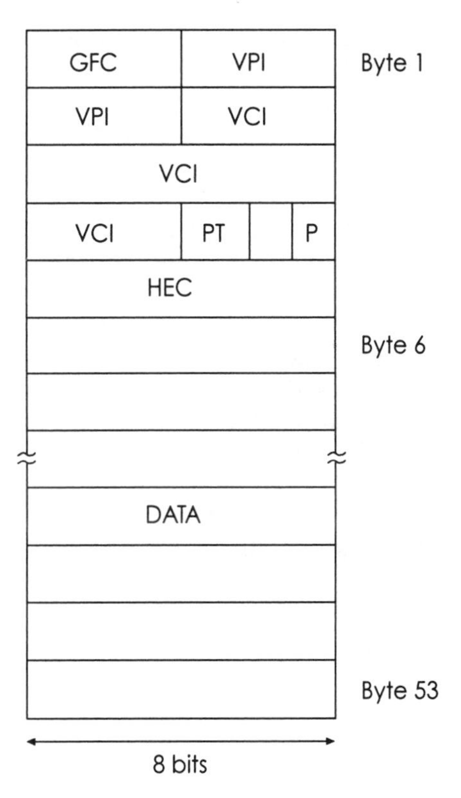

Figure 5.14 ATM Packet or cell

The system has an interesting variation on error control coding. The packet data is not checked or protected in any way. This betrays the origins of ATM in telecommunications, where 'drop outs' or lost packets in telephony are not critical. For many data applications, data loss is significant and in those cases the designers of ATM expect the user to provide error detection or correction at a higher level. This just means that if a user wants the facility it must be provided within the data field. This is an interesting contrast with X25, where error detection is part of both the frame and the packet level operations. The header field is provided with a single byte checksum to ensure packets are not routed incorrectly. A secondary function of the checksum is to act as a delimiter between the header and the data. ATM is pared down to have minimal overheads since there are no packet opening or closing flags. Nodes receiving or dealing with ATM packets simply identify the checksum and count bytes to define the packet limits. The data field is

subjected to a scrambling algorithm to ensure that no false or imitation checksums are found. The purpose of the first five bytes is as follows:
Byte 1: first nybble (4 bits) – GFC. GFC is generic flow control, which is for future development and serves no useful purpose.
Byte 1: second nybble – VPI. This is the virtual path identity
Byte 2: first nybble – VPI continued.
Byte 2:, second nybble – VCI. The virtual circuit identity.
Byte 3: and first nybble of Byte 4 – VCI continued.
Byte 4: second nybble – PT and P. PT defines the payload type and P is the bit that sets the cell loss priority. If this bit is set, switches are at greater liberty to discard the packet if they need to.
Byte 5: HEC. – the header checksum.
The purpose of virtual path identity and virtual channel identifier need explanation. The telecommunications origins of the system mean that the preferred method of operation is by virtual call. This connection oriented style is compatible with X25 and telephone principles. As with X25 the destination address is hierarchical in terms of virtual group and virtual channel. Because the calls are set up between ATM switches the hierarchy is expressed in terms of 'paths' and 'circuits' within those paths. This switch to switch assumption allows the further assumption that packets belonging to the same connection will travel along the same path one after another and so will always remain in sequence. Virtual circuits are set up when a call is established. Figure 5.15 illustrates the principles behind virtual paths and circuits.

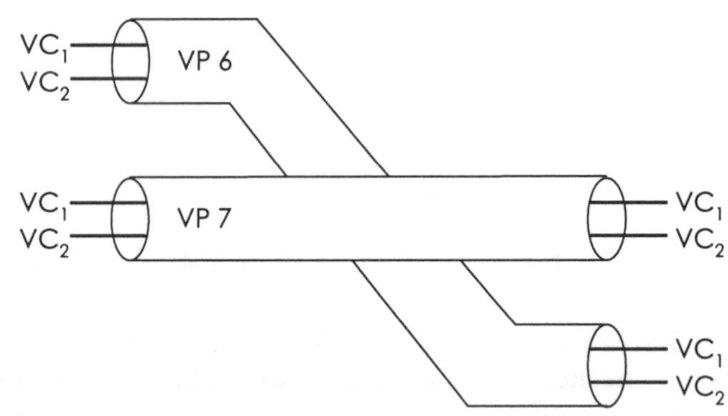

Figue 5.15 Virtual paths and virtual circuits

5.8 Routing

In packet switched systems the sources and destinations of data, whether they are part of a local area network or not, are connected together by a series of data highways. The geographical diversity and distribution of possible sources or destinations mean that it makes sense to create an irregular network in which some network nodes or routers may act as forwarding devices. The precise nature of the forwarding process will depend on many things, amongst which are: the number of alternative routes to a particular destination; the necessary robustness in terms of the response to topological change; variations in traffic densities offered; the fairness in terms of allowing nodes sufficient access; and stability or ability to recover from highly congested and backlogged states. There are two representations of interconnecting networks. The simplest is the irregular network diagram showing the nodal points and the paths connecting them. See Figure 5.16 for such a diagram.

Whilst many interconnection paths are possible it is clear that optimal paths exist. To understand this, consider nodes T and Z in Figure 5.16. It appears possible to reach Z from T via V,W, and X. The optimal path is the one that uses the fewest relays and in this case would be T,Y,Z. Just one intermediary is involved, compared with three for the route T,V,W,X,Z. This fact provides the basis for the second representation for an irregular network. The system could be drawn using only optimal routes and eliminating all others. This kind of diagram is called 'sink tree'. Figure 5.17 shows a possible sink tree for the network of Figure 5.16.

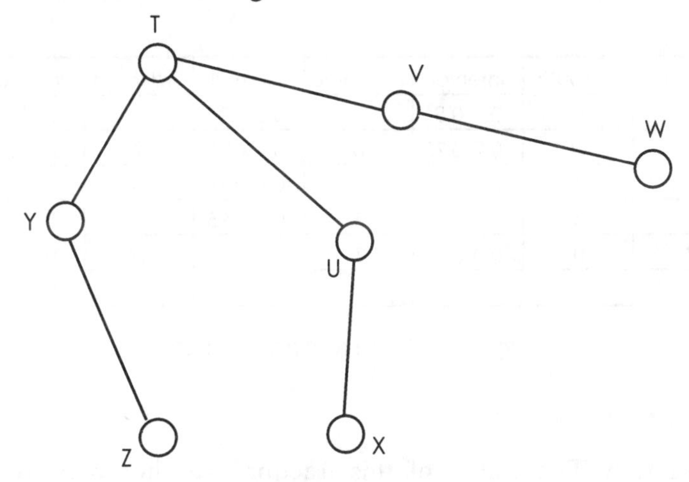

Figure 5.17 Sink tree

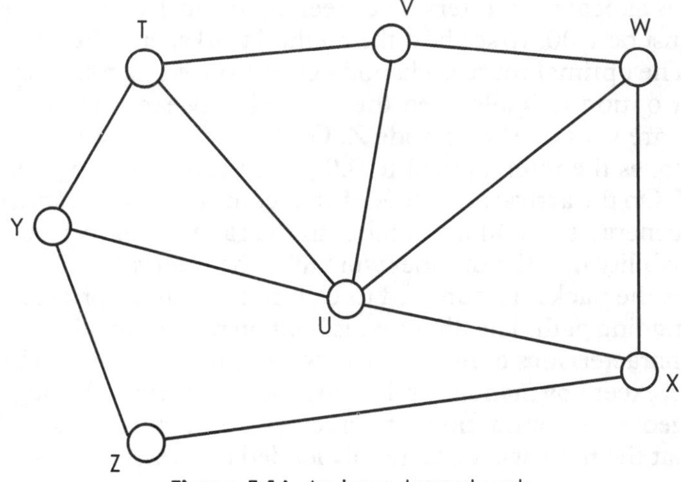

Figure 5.16 An irregular network

Optimal routes can be chosen on the basis of several possible criteria. The number of relays is one but an alternative is to determine the shortest packet delivery time routes. These optimal routes may or may not be the same. The optimal path may also change if it is based on delivery time and the traffic offered causes congestion on a particular route. Adaption is seen as an advantage in cases such as these but adaption adds complexity.

5.8.1 Directory routing

Directory routing is a non adaptive method. Each node carries a look up table which has been pre-loaded by the network management staff. The look up table is a directory, showing the onward routing paths for a given destination. The format of a look up table is shown in Figure 5.18. The table shown is that belonging to node X in Figure 5.16. Note that three options are given for forwarding a given packet. Associated with each possible outgoing route is a

Destination	1st route	Interval	2nd route	Interval	3rd route	Interval
U	U	0.0–0.8	W	0.8–0.95	Z	0.95–1
W	W	0.0–0.75	U	0.75–0.9	Z	0.9–1
Z	Z	0.0–0.6	U	0.6–0.85	W	0.85–1
Y	U	0.0–0.9	Z	0.9–0.95	W	0.95–1
T	U	0.0–0.8	Z	0.8–0.9	W	0.9–1
V	W	0.0–0.7	U	0.7–0.95	Z	0.95–1

Figure 5.18 The routing directory

decimal fraction. The values of this fraction are chosen by the network managers to reflect the optimality of a route according to chosen criteria. Each route is allocated an interval between 0.00 and 1.00 and the sum of the intervals must be 1.00. To see how the method works, take the possible routes to node T. The optimal route is via node U and since this route represents by far the best option it is allocated the interval between 0.00 and 0.75. The alternatives are via node W or node Z. Of these W is considered marginally better and takes the interval 0.75 to 0.9, whilst the remaining 0.9 to 1.00 is owned by Z. On the arrival at node X of an incoming packet destined for node T, node X generates a random number in the range 0.00 to 1.00. There is a strong probability that the number will fall in the interval 0.00 to 0.75 and if this happens the packet is launched to U. If not, the appropriate interval will select the ongoing path. The directory is built on mean statistical data and the burst like characteristics of many networks may justify adaption but it is just not possible, except by manually reloading the directories. Although adaption is advantageous the attraction of directory routing is its simplicity and provided that the network is not heavily loaded or stressed the results can be good.

5.8.2 Adaptive routing strategies

The general idea is that in an adaptive system the nodes would be able to monitor traffic densities and flow in order to make adjustments to the routing applied to a given packet. The notion is straightforward but the implementation is not. The simplest option is to have a central control node whose purpose is to gather data about how traffic is flowing and respond by adapting the routing tables of all the other nodes. The problem is that in order to monitor activities the central control node must receive regular updates on individual node performances. Just this control traffic can provide a significant loading, especially when control messages funnel in to the area of the network occupied by the control node. The overhead of control packets both going to the control node and going from the node in order to update a remote node routing table can make centralised control unattractive.

A number of possibilities for devolving control to ordinary nodes have been created. The simplest merely has each node monitor the queue size for traffic leaving for other nodes. The decision about how to forward an incoming packet may be taken on the basis of route choices, weighting intervals and queue length. If the packet would normally leave by a certain pathway but there is evidence of congestion leading to delay then the node may attach the packet to the shortest outgoing queue path.

Another possibility is to create the conditions whereby nodes can 'learn' about the network performance by reading a key field in packets passing. To facilitate this the special field usually contains a counter which is incremented each time the packet transits a node. The data packet naturally carries the source node address of its message and a node may 'build up a picture' of the network by observation and analysis. It will be able to calculate the number of intermediate nodes between itself and the source address even though the packet may be destined for a different node. If problems arise on the network, packets from a given source may arrive by different paths, perhaps with more relays, and the incoming path might help in identifying the likely site of trouble. Once the position of a trouble spot has been deduced a node may adapt its own routing table to avoid sending more packets through a clearly stressed zone.

Those working on these problems soon pointed out that, unless controlled, the results of adaption by individual nodes could result in isolated parts of the network being run down and ultimately ignored. The suggestion was to allow adaption but to force nodes to periodically forget all they had learned, look at the network again with an 'open mind' and build a new picture. Other options include a form of hybrid operation where each node is able to measure the relative performance of the links it is connected to but has to report its findings occasionally to a central node which is programmed to detect anomolies and dictate new commands which keep the situation under control.

5.9 Interconnection and the case study

The four 10base5 Ethernet segments in the headquarters building are best interconnected by repeaters but what about extending the network outward to branch offices? The material covered in this chapter does allow interconnection if the distances are up to 100 km. To evaluate the suitability of links it is necessary to measure the likely traffic load and characteristics. The nature of the business involved in the case study means that it is likely that so called branch offices may be of two possible kinds: regional operations centres and insurance/motoring shops. The latter are more likely to generate sporadic data whilst the former will consistently produce large numbers of messages and database accesses. Traffic growth is endemic to data networks and it is important that sufficient capacity is provided for. The range of FDDI, 100 km (62 miles), commends its use for many of the centres in question and the 100 Mb per second data rates would be adequate for this backbone operation. FDDI is relatively mature and ATM is not. A number of critical design features of ATM, particularly relating to the switches, have yet to fully emerge from the research and development laboratories. Reliability is a key requirement in a commercial application and consistent, known performance is also important. More distant centres may have to be connected using services offered by the public data network.

Questions

1 Compare the functionality of repeaters, bridges and gateways and comment on the usefulness of a multiport repeater in an application such as the case study head office building.

2 Discuss the applications for which FDDI was designed and differentiate between class A and class B stations or nodes.

3 X25 is an access protocol to the public data network. Define the scope of the standard and contrast the system with that expected for the Internet.

4 Congestion control is one of the main problems in an ATM network. Explain why this is so and rehearse the mechanism used by ATM to deal with the difficulty.

5 Two popular routing strategies are directory routing and adaptive routing. Analyse the suitability of these options for the case study application.

6

Going further

6.1 Background

The specification in the case study expects to extend the facilities originally available at the headquarters building to branch offices. This will mean using the public carriers of communications. The variety of services offered by these carriers requires choices to be made and for these choices to be made appropriately it is necessary to understand both the technology offered and the marketing options.

Communications networks have a long history, developed since the invention of the telephone. In recent times there has been a strong trend to use digital signalling because of the role played by computers in providing advanced features. The original telephone system carried speech as an analogue signal which was band limited to a range from 300 Hz to 2.7 kHz. The systems performance characteristics are shown in Figure 6.1. The two curves represent the performance of the services normally available, The public switched telephone network (PSTN) and leased lines. The latter

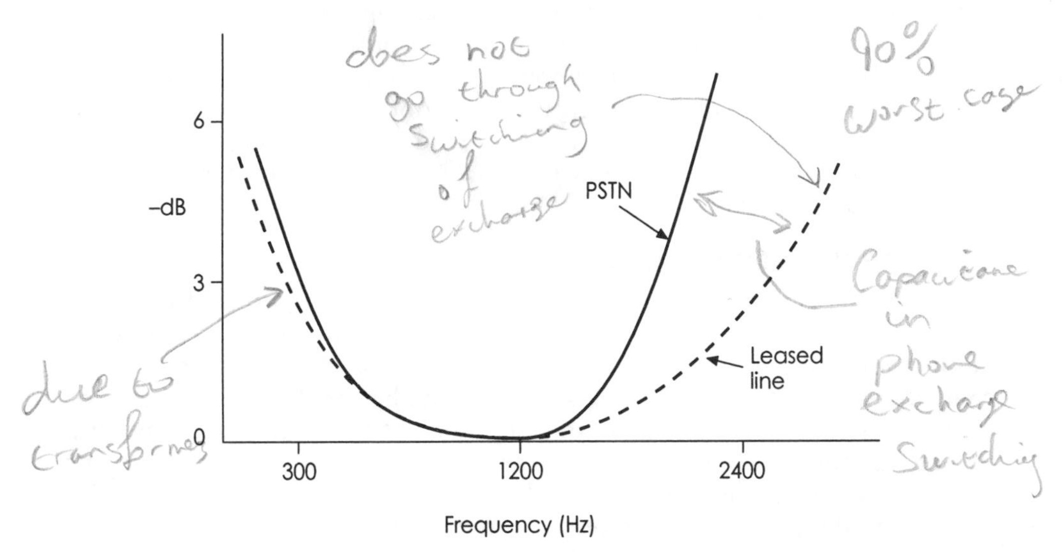

Figure 6.1 Telephone system characteristics

provide a permanent connection between two points, whilst the former allows 'dial up' connection through the use of a switching frame in an exchange. Although many exchanges are now digital as are many long distance circuits, the local lines to subscribers remain analogue. Analogue to digital conversion takes place at the exchange. The telephone system user thus 'sees' an analogue system at his end. Using the telephone system for data transmission requires an understanding of the inherent constraints.

Public carriers also offer a digital interface for data other than standard telephone traffic. This entails providing a subscriber with a digital line and access to the public data network (PDN). There is no difference in the line, only in the way it expected to be terminated both at the exchange and with the user. Obviously, if the data arriving at the exchange is already digital it does not need analogue to digital conversion but must comply with a defined access protocol. Inevitably all lines in the future will become fully digital but this assumes an 'integrated service'. At present it is necessary to consider all three possibilities:

- the use of analogue circuits for data transmission
- use of the public data network
- integrated services digital network (ISDN).

6.2 Analogue data signalling

The real problem with signalling over telephone circuits is the bandwidth limitation. It might be asked why digital 0s and 1s cannot be signalled directly. The section on harmonic synthesis in Chapter 2 explained that a rectangular wave train has a fundamental frequency equal to the reciprocal of two times the bit period. A number of harmonics must be available in order to form a reasonable 'square' shape. Typically, all components up to the fifth harmonic are needed. Taking an arbitrary signalling rate of 1200 bits per second, resulting in a fundamental of 600 Hz, it is clear that insufficient bandwidth is available in the telephone system. Whilst matters ought to be improved by reducing the signalling rate it cannot be reduced below 600 bits per second (fundamental = 300 Hz) before moving below the lowest passed frequency. This unsatisfactory state of affairs leads to the consideration of the role modulation might play.

6.2.1 The use of modulation

Of course there are many forms of modulation, such as amplitude, frequency, phase and pulse coded. First, consider frequency modulation. In frequency modulation the frequency of the carrier sinusoid is changed to follow the characteristics of the modulating signal. Normally the degree of frequency change or shift is proportional to the amplitude of the modulating wave and the rate of change of frequency of the carrier is proportional to the frequency of the modulating wave. If the modulating signal is data, i.e. a rectangular pulse train, then the amplitude only assumes two possible values (for 0 and 1) and the frequency of the modulating wave is the underlying fundamental component. A specific carrier frequency change may be set between a 0 and a 1. Say this is 200 Hz. Further assume that the data rate was 1200 bits per second (fundamental 600 Hz). The bandwidth of a frequency modulated signal may be estimated as follows:

$$\text{Bandwidth} = 2(F_d + F_m)$$

where F_d is 200 Hz and F_m is 600 Hz.

If the 0 and 1 values for the carrier are 1080 Hz and 1280 Hz, then the bandwidth will be 1600 Hz centred on 1180 Hz. The signal will be accommodated between 380 Hz and 1980 Hz, well within the limits provided by a telephone circuit. The spectrum diagram for this example is given in Figure 6.2.

Whereas 1200 bits per second was not possible over a telephone circuit with a raw binary waveform, it certainly is possible when the binary states are signalled as two different tones. Obviously a conversion box is necessary when connecting a computer serial port to a telephone line. The conversion actually performs a modulating function and clearly, to receive information would require a demodulator also in the box. The combined two way converter is called a modulator – demodulator or modem.

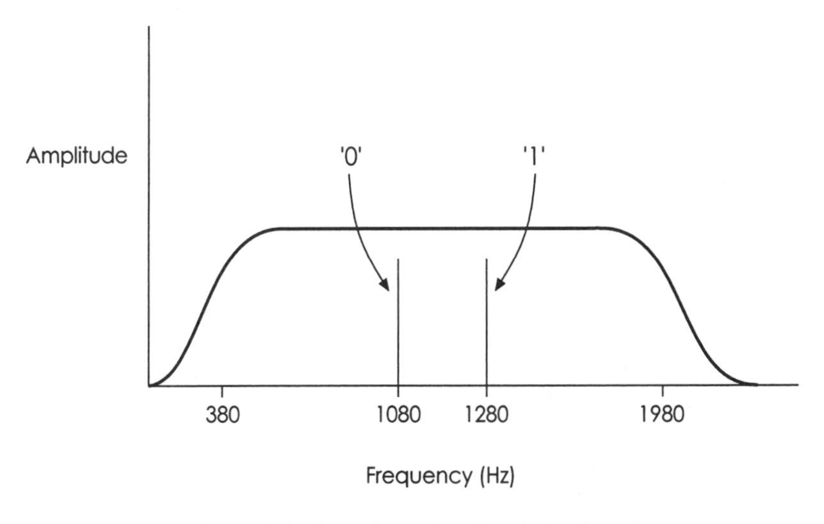

Figure 6.2 Spectrum for FM data signal

6.2.2 CCITT V21 modem standard

The V21 standard defines a low speed system using frequency modulation which allows two way simultaneous transmission at 300 bits per second.

The 0 and 1 tones for each direction are separated in the spectrum but because the difference in tone frequency is 200 Hz in each case and because the fundamental frequency of the signalling rate is 150 Hz, each direction accounts for a bandwidth of 700 Hz. One direction occupies the approximate range 700 – 1400 Hz, whilst the other uses 1400 – 2100 Hz. Signalling is thus frequency division multiplexed. Figure 6.3 shows the channel allocation and Figure 6.4 the simplified system. In Figure 6.3 the upper channel is always used by the modem that is being 'called' by another. The calling or 'originating' modem takes the lower channel.

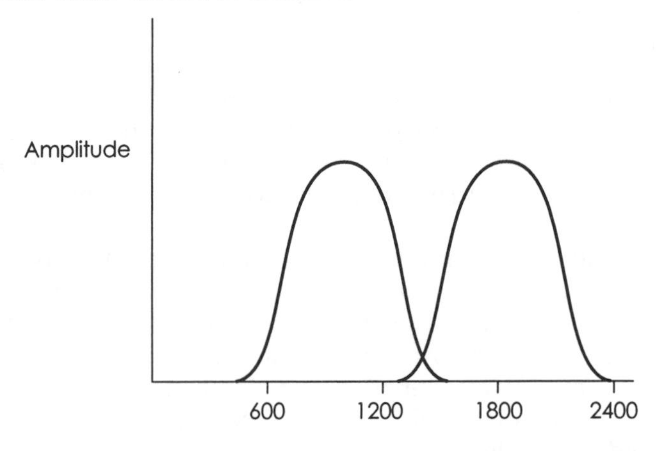

Figure 6.3 Spectrum for the V.21 modem standard

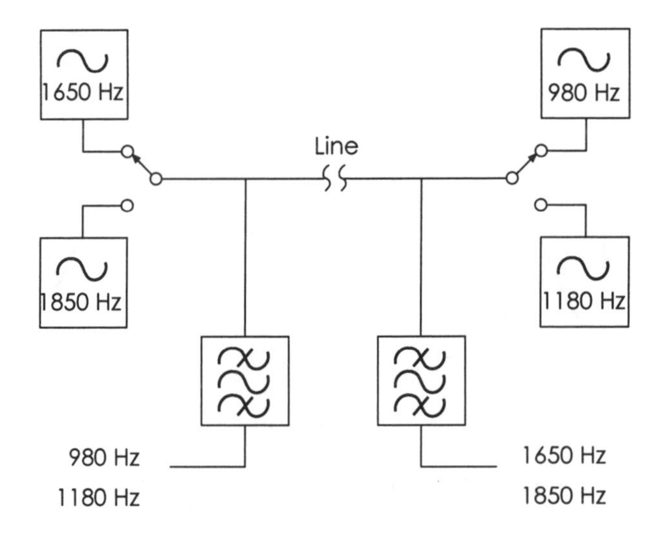

Figure 6.4 Simplified modem system

6.2.3 CCITT V23 modem standard

The V23 standard uses the available bandwidth in a different way. It is designed for use when assymetrical device speeds apply at each end. A good example of this is the system used by travel agents which expects agents to dial up the computer systems of specific holiday companies in order to receive colour screens of block (Teletext style) graphics and to select bookings and information by minimal key press returns. It would be useful to refresh the screens as quickly as possible and so the majority of the bandwidth is allocated to this. Occasional key presses need very little bandwidth.

The tone frequency difference in this case is 800 Hz and the signalling rate of 1200 bits per second, results in the main channel occupying 500 – 2800 Hz. A 75 bits per second return or 'back' channel is located just below the main channel, as illustrated by Figure 6.5.

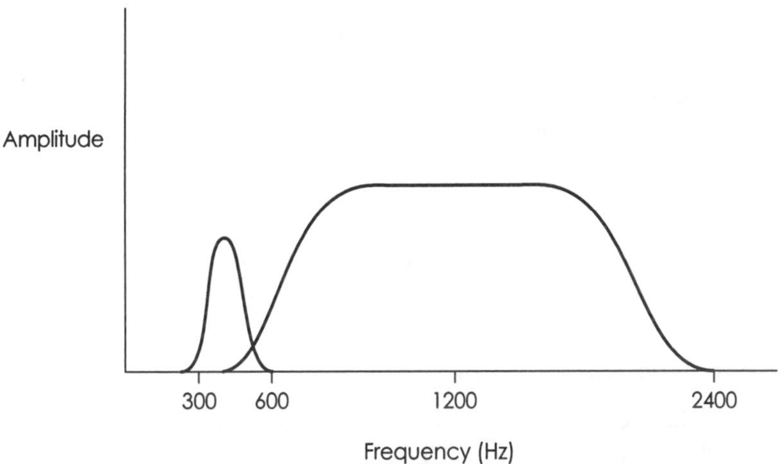

Figure 6.5 Spectrum for the V.23 modem standard

6.2.4 The CCITT V22 modem standard

If a modem could signal 0 and 1 states without changing frequency, then potentially less bandwidth might be used. One option is to alter the phase of the signal instead. Another big advantage accrues if the data rate is divorced from the signalling rate. To appreciate this, consider a system capable of signalling four different phase shifts. Say the signal could be made to suffer a 45, 135, 225 or 315 degree phase change. Each of the four possibilities might be associated with a pair of binary digits. These dibits, 00, 01, 11, 10 allow a byte to be signalled by just four signalling states. This system is used by V22 and it is called quadrature phase modulation. Because a dial up telephone link can really only accommodate about 1200 signalled states per second, V22 can signal 2400 bits per second. This represents about the limit allowed by some countries for modem signalling over their public switched telephone networks. To achieve greater data rates it is necessary to lease a private line. Private lines are 'wired through' exchanges and thus may only provide a fixed link which is generally paid for on an annual basis. Because leased lines are not connected to crossbar switching, there is less of a problem caused by stray shunt capacitances which act as low pass filters. Also crosstalk between circuits is reduced; and an extension of the upper bandwidth limit and less interference are characteristic of leased lines. Figure 6.1 illustrates the comparison.

6.2.5 Leased line modems

A natural development of quadrature phase modulation was to extend the number of discrete phases that are signalled. An eight phase system may signal tribits of information with a consequent increase in maximum data rate. It is quite difficult to build phase shift detectors capable of reliably differentiating between more than eight phases and to signal even faster requires a composite modulating scheme. The 9600 bits per second V29 modem uses eight phases and four signal amplitudes, whilst the V32 modem uses twelve phases and three amplitudes. Both yield 16 signal states and are able to signal 'quadbits'. A byte is thus signalled in just two signalling states. The relationship between the number of different symbols a modem can produce and the absolute maximum data rate that it can generate when being used over a link of given bandwidth is as follows:

bits per second = bandwidth × number of bits per symbol

This equation is derived from information theory developed by Shannon and for V29 or V32 the calculation produces a result of about 10 000 bits per second, assuming a line bandwidth of around 2.5 kHz. The figure of 9600 bits per second is thus close to the limit achievable.

6.2.6 CCITT V29 modem standard

To detect the differential phase shifts in the signalling a reference is needed. The local reference is provided by a voltage controlled oscillator in the receiving modem. When communication between two modems is initiated, the sender starts by transmitting a 'preamble'. The preamble is fed to a 'clock recovery' circuit as shown in Figure 6.6(a).

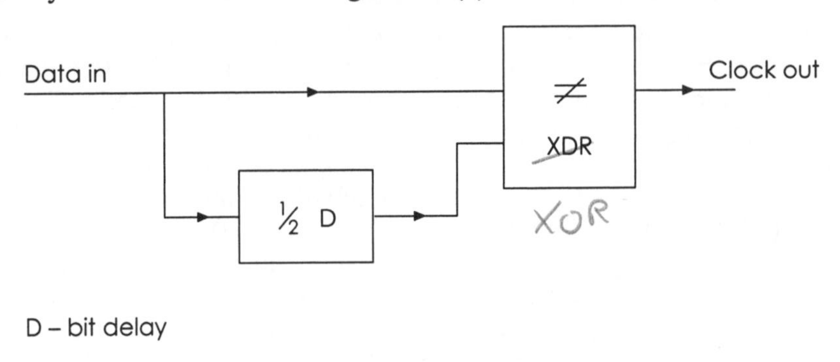

Figure 6.6 (a) V.29 clock recovery

The preamble is a signalling pattern which contains four 'segments' but includes two kinds of alternations. The first is for synchronising purposes, whilst the second is an equaliser conditioning pattern. After detection the synchronising signal produces a 01110101 string. The clock recovery circuit exclusively ORs this string and a half bit delayed version of itself. Figure 6.6(b) shows the resulting clock which is used to phase lock the reference

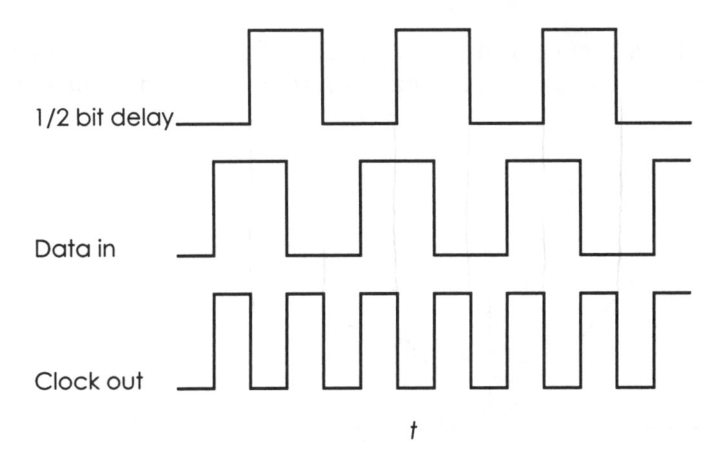

Figure 6.6 (b) Clock recovery waveforms

oscillator. Once the latter is locked, accurate phase comparisons can be made. The repetitive pattern is signalled as large phase shifts that are relatively unmistakeable and the pattern allows crude demodulation even with the free running reference oscillator. When 'in lock', the reference enables the differentiation of message signal patterns which can be only 45 degrees apart. The full range of possible signalling states is illustrated by Figure 6.7, which is called a 'constellation' diagram. Each point represents a

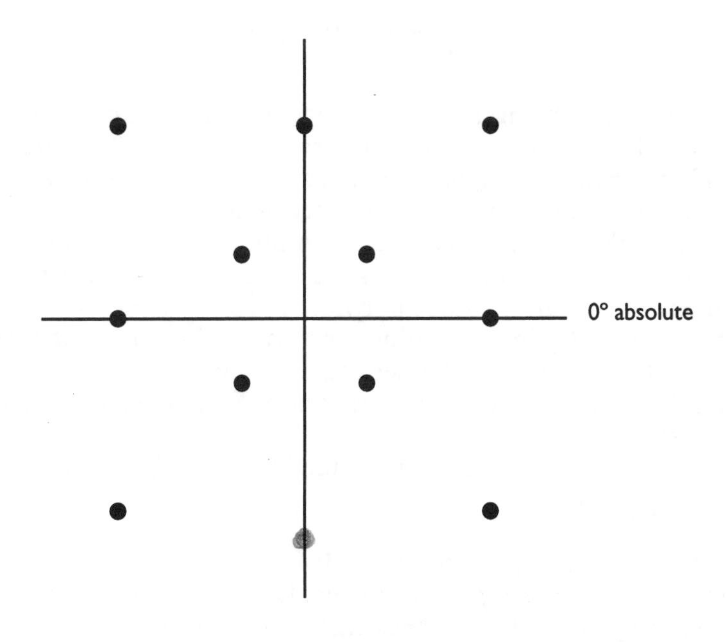

After: CCITT 34530,V29

Figure 6.7 V.29 constellation diagram

signalling state for a particular quadbit pattern. Figure 6.8 shows the phase and amplitude differences between points on the constellation diagram.

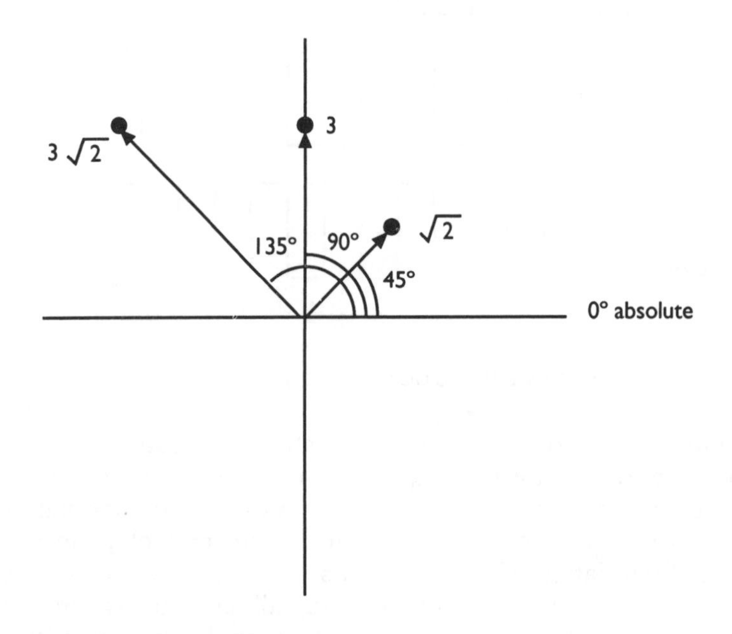

Figure 6.8 V.29 vectors

Once a message is being signalled it becomes important to maintain the reference oscillator 'in lock'. A sequence of 1111 or 0000 quadbit signals would, after demodulation, provide no useful synchronising data. To avoid a loss of reference and consequent erroneous demodulation a special technique is used. The method is called scrambling and the purpose is never to allow long runs of constant state to cause loss of lock. At the transmitting end a scrambler divides the message polynomial (see Chapter 4, cyclic, systematic codes and CRC) by generating a polynomial to create the transmitted sequence. At the receiver, the sequence is multiplied by the generating polynomial to recover the original message. Scrambler logic is defined in Figures 6.9(a) and (b).

In Figure 6.9(a), the message is divided by the polynomial:

$$1 + x^{-18} + x^{-23}$$

The coefficients of the quotient of this division are taken in descending order from the data sequence to be transmitted. During the initialising synchronising segment sequences, the '0' input ensures the correct starting sequence. D_i being the input message and D_s the scrambled output.
For descrambling at the receiver in Figure 6.9(b), D_0 is the recovered message.

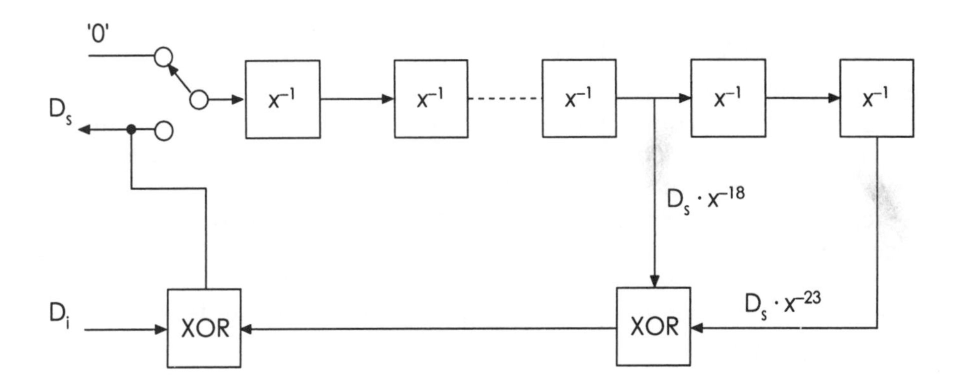

D_s – scrambled data
D_i – input data
x – bit delay

After : CCITT 34552, V29

Figure 6.9 (a) Scrambler

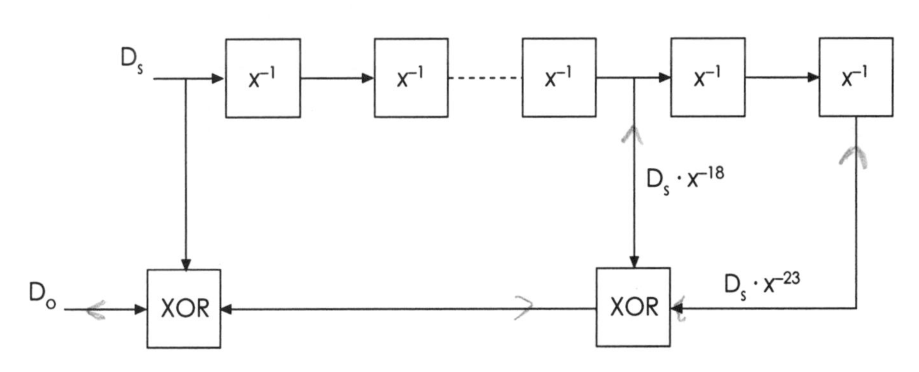

D_s – scrambled data
D_i – data out
x – bit delay

After : CCITT 34562, V29

Figure 6.9 (b) Descrambler

6.2.6 CCITT V32 modem standard

Although V32 is being discussed as a leased line system it was also designed to operate over the general switched telephone network. For this reason, although it shares a number of similarities in principle with V29, it differs in a number of significant details. The basic form of differential phase and amplitude modulation is retained but signalling of quadbits is now achieved.

6.3 The public data network

The public data network (PDN) is vast set of data switched exchanges (DSE) and high speed links running at 600 Mb per second and more. The internal protocols of the network are invisible to the users but the network does have a series of 'access protocols' with which users must comply in order to interface properly. Since company private networks may need to use carriers' services for certain long and medium haul links, it is these access protocols that will be emphasised. Two main classes of service are offered: virtual circuits; and packet switching. These options are marketed under names such as Kilostream and Megastream for the former and the packet switched stream (PSS) for the latter. Because the standards describing these services are described by numbers prefixed by 'X' the whole range of offerings is referred to as the X-Stream.

6.3.1 Kilostream, Megastream and X21

These are the virtual permanent link access standards. The user rents or leases a digital line from his own premises to the DSE. At the user's end the line is connected to a network terminating unit (NTU), which is a modem equivalent but which provides binary modulation and synchronous signalling according to the WAL2 format. Figure 6.10 illustrates the scheme

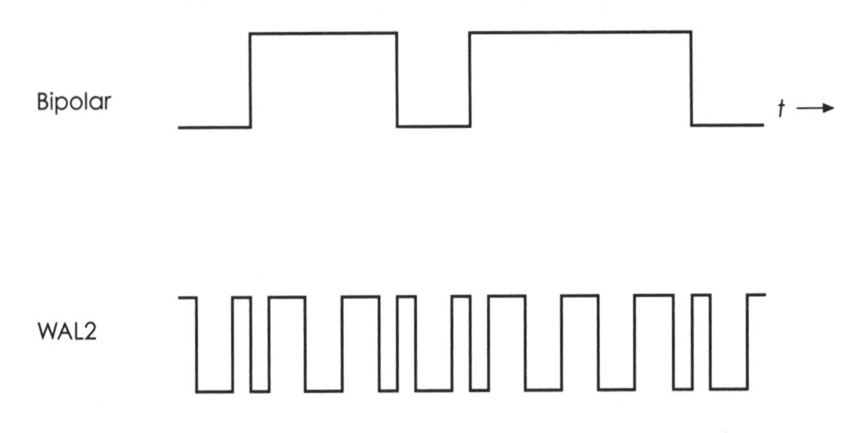

Figure 6.10 WAL2 format

The NTU expects the user's computer or terminal to communicate using the procedures and signalling defined by X21. The standard describes an interface between DTE and a DCE. These abbreviations describe the NTU and the user's terminal. The data circuit is considered to be the digital line link between the user's property and the DSE. Thus the NTU is a 'data circuit terminating equipment' or DCE. A data terminal is a DTE. Figure 6.11 shows the scope of X21 in a Kilostream/Megastream connection. Kilostream NTUs allow for signalling at 9600 bits per second (bps)or 64 k bps. Megastream provides 1Mb per second and at the DSE an incoming digital line is connected to an appropriate data rate tributary card which feeds the

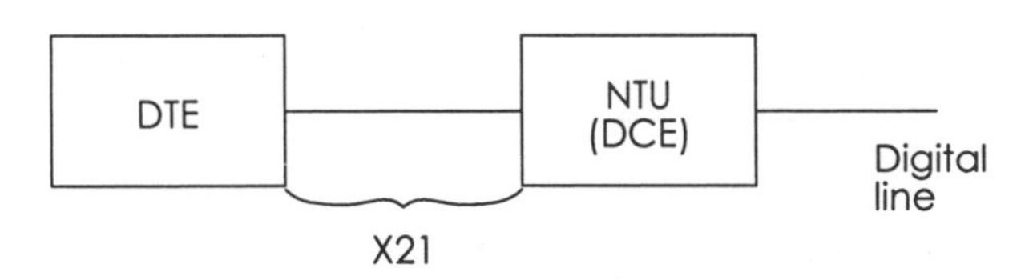

Figure 6.11 The scope of X.21

signal into the internal protocols of the PDN. Figure 6.12 shows tributary and crossover sites for a 9600 bps Kilostream connection between two users' digital lines in different parts of the host country. The important point to realise is that the two end users' terminals may talk to each other at any time and no addressing needs to be done. It is as if a permanent link exists between them, although, of course, time division multiplexing is carried out on the internal routes within the public data network.

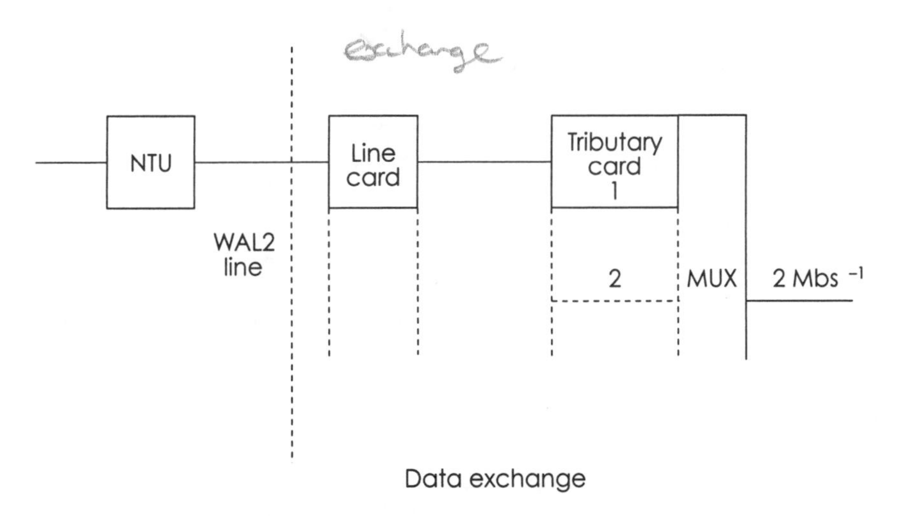

Figure 6.12 Tributary connection

The seven circuits of the X21 interface: receive; transmit; control; indication; element timing; signal ground; and common return operate in a precise way when a link is to be activated. Ready states mean that DCE and DTE may exchange data. Controlled not-ready implies that although there is no fault there is a good reason why communication is not possible. Typical of these good reasons is that the DTE or DCE may have just been switched on and are busy initialising themselves. Uncontrolled not-ready signifies a fault.

Activating the circuit is achieved in a way that is analogous to the stages in making a telephone call and Table 6.1 illustrates the steps. Step 0 is the idle state, which will persist for as long as the link is inactive. Assuming a DTE activates the link, step 1 is achieved by asserting logic 0 on the transmit circuit. On recognising this change, the DCE replies by issuing an ASCII code string '++++...'. This sequence is analogous to the dialling tone when a telephone call is made. The DTE is expected to respond by sending the

Step	C	I	Event	DTE sends (T)	DCE sends (R)
0	off	off	line idle	1	1
1	on	off	DTE initials	0	
2	on	off	DCE responds		+++++...
3	on	off	DTE dials	address	
4	on	off	remote reached		call progress
5	on	on	remote responds		1
6	on	on	dialogue	data	data
7	off	on	DTE closes	0	
8	off	off	DCE closes		0
9	off	off	DCE hangs up		1
10	off	off	DTE hangs up	1	

Table 6.1 Illustration of the steps

address or number of the called user. During subsequent moments, the DCE returns code messages from the public data network concerning the progress being made in securing a connection. Subsequently the called user's DTE responds and a conversational phase of any length ensues. Eventually it might be that the caller wishes to 'clear down' the call. Asserting logic 0 on the transmit circuit and steps 9 and 10 finish the sequence.

The overall analogy is of leased telephone line operations but only digital data is exchanged during the conversational phase. The signalling is synchronous, but no clock recovery or binary modulation are present since a separate sampling synchronising signal is carried by the timing signal element circuit. Data is transmitted in a 6 + 2 mini frame format such as is described by Figure 6.13. Generally charges for this kind of link are based on an annual fee which would be much greater for the Megastream service than for the Kilostream options. Many organisations may have existing DTEs which are not X21 compatible but are compatible with earlier interface standards such as V24. Such equipment can connect to the NTU (DCE) provided that the NTU is specified as an X21bis device. X21bis refers to a variant of X21 and it is simply V24. Because of the age of the V24 standard, it is incapable of the more sophisticated signalling such as the call progress messages and their

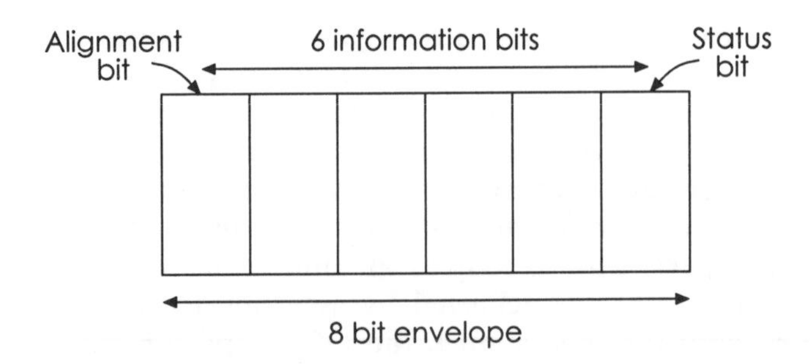

Figure 6.13 X.21 mini frame

interpretation. X21bis is intended to provide an easy and gradual upgrade path by being a kind of intermediate step to full X21.

6.3.1 X25 – the packet switched stream

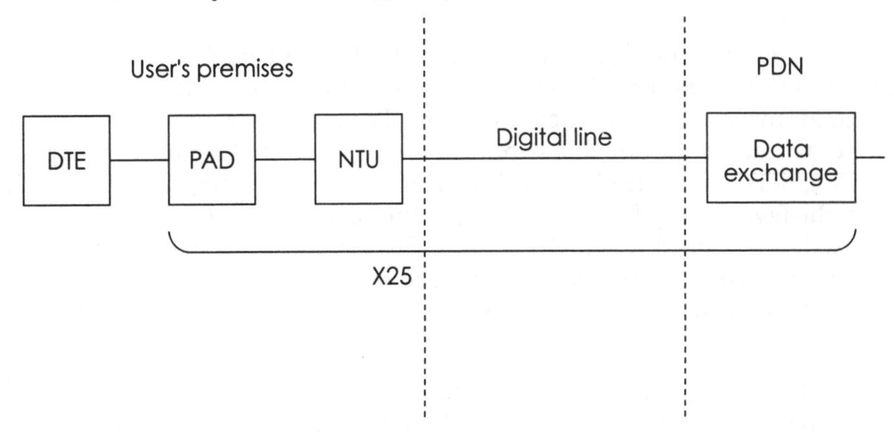

Figure 6.14 Scope of X.25

X25 is the standard that defines the 'access' protocol to the packet switched stream. The ISO 7 layer model is pre-dated by X25 but there is remarkable similarity up to the limit of level 3. The public data network (PDN) is not concerned with any issues addressed by higher layers since it seeks only to provide users with links over which they can do as they like. It is thus their responsibility to decide on session control and the applicability or otherwise of data encryption. The scope of X25 is described by Figure 6.14 which makes it clear that the purpose is to connect a user with the nearest digital switched exchange (DSE).

Layer 1 is specified as using X21 or X21bis, but layer 1 is only relevent between the DTE and DCE. Overarching the DTE/DCE link and the digital line to the exchange is the the link layer, layer 2. This is a data frame based operation which provides error and flow control based on a variant of synchronous data link control (SDLC) called link access protocol- B (LAPB). LAPB is bit oriented and uses a frame format as illustrated by Figure 6.15. Whereas SDLC was created by IBM to provide a multi-drop environment, LAPB is invariably used as an end to end system. The frame operation allows for error detection by cyclic redundancy check (CRC 16 bits) and in the

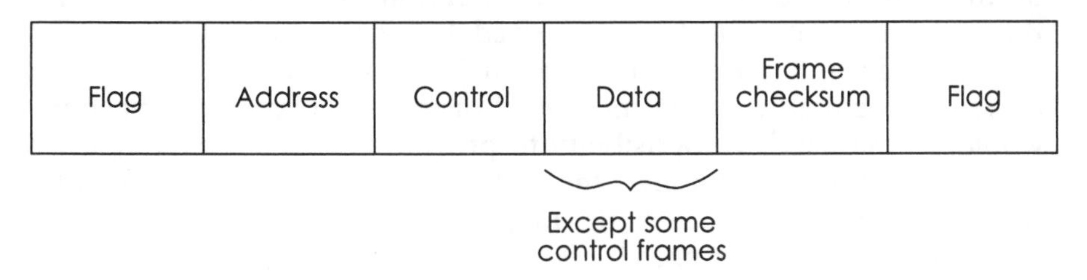

Figure 6.15 LAPB frame

event of an error being detected the correction strategy is 'Go Back *n*'. It must be realised that this error detection only applies between the exchange and the user and not over the public data network to some remote correspondent. It is not an end to end error detection and correction system at the frame level.

Nested within the LAPB data frame is the X25 packet, complete with header and user data fields. The packet operation is a virtual circuit system. If datagrams were used, each would be launched independently into the public data network but this is not the method used. As might be expected from carriers whose traditional business has always been telecommunications, the favoured style is the virtual circuit. Consequently, a call must be set up, used, and closed down. To achieve this a number of packet formats are necessary. Figure 5.7 illustrates the three basic types.

The detailed sequence of events involved in an X25 communication is as follows:

1. A call request packet is sent to a destination by the originating node. The format of this packet is illustrated by Figure 5.7(a). An item of special interest is the 'channel' field which betrays the fact that an X25 node may multiplex packets for different destinations. When a connection is ordered the address of the destination node must be supplied. The originating node allocates an arbitrary channel number to this 'conversation'. Once the data exchange associates a particular channel with a particular destination address it is remembered until the conversation is cleared. This explains why only the call request packet needs to carry the destination or 'called' address. The exchange is able to make a number of such associations and thus the originating node may set up a number of concurrent calls. The packets for these calls are sent to the exchange over the same digital line and the exchange steers the data to the appropriate destination. A channel number may be any number between 1 and 4096. Channel number 0 is reserved for future enhancement. The control bit simply differentiates between control and data packets: 0 for data, 1 for control. The call request packet is one type of control packet.
2. The exchange tries to connect to the remote destination and if successful returns a 'call accepted' packet which is also a control type but with the simpler format shown in Figure 5.7(b).
3. An exchange of data packets will then take place until the call is cleared by either party. The data packet format of Figure 5.7(c) does not carry the destination address, but only the channel number. The exchange relates this to the destination address it received during the call request sequence. The Q field is intended for use as a 'qualifier'. Because X25 is a network access protocol only,there is no layer 4 or transport layer specification. It would be the user's responsibility to provide the layer functionality for layers 4 and above. In doing so, the user might use the Q bit to differentiate between control and data entities in his transport protocol. The sequence and piggyback fields are used for flow control and the modulo field defines the window size (01 = 8,10 = 128). Window sizes define the range of

sequence numbers for the packets. The piggyback field has two possible uses. When the D bit is 0, any acknowledgments are only from the local DCE. When D is 1 then the acknowledgment is end to end on the link. Making D = 0 inplies that delivery at a remote destination is not guaranteed. The 'more' field allows groups of packets to be associated, so that the system knows that they belong together.

4. Eventually one of the channel ends wishes to terminate the conversation. It generates a 'clear request', which will be followed by a 'clear confirmation' from its remote partner. The call is thus cleared down and the channel number would be available for another user.

The call request packet carries the source and destination addresses as decimal numbers and since addresses may have varying lengths it is necessary to cope with any eventuality by indicating the actual length of each address. In a similar way the 'facilities' length field simply states how many bytes of Facilities information follows. Facilities are services offered by the carrier. Typical examples are: reverse charges; priority delivery; or closed user group use. There are minor variations between US and European carriers and so the facilities field may also be used to signal the maximum packet size in bytes or the window size for the sliding window protocol. If this signal is not provided European systems default to 128 bytes and 2 respectively. The user data field could be used to indicate what process in the recipient node peripheral is to be communicated with.

Control packets are often header only entities but sometimes additional information is appended. For example a clear request packet may also carry additional information about why the call needs to be cleared.

6.3.2 The X25 PAD

A PAD is a packet assembler disassembler. Whilst some equipment such as host computers and terminals may contain all the necessary features to allow them to generate and receive packets there are many devices that are unsuitable. A typical example is that of a device which operates asynchronously such as a PC working through its COM1 or COM2 serial ports. Other equipment such as VT series terminals also cannot directly operate the X25 protocol. To connect these systems to the public data network an interface box is needed. The box may be at the network user's premises and will need to be connected to a digital line via an X21 network terminating unit (NTU). Figure 6.16 illustrates the arrangement. The PAD acts like a form of concentrator, assembling individual characters or other entities into X25 data packets. Of course it also performs the inverse operation for data flows in the reverse direction.

In some cases a user might be sufficiently isolated from the public digital network that no line is available. In these cases an analogue telephone line may be pressed into service. A V32 modem would be necessary to achieve equivalent performance to an X21 link and the question arises as to the location of a PAD in this scenario. Normally PADs are placed at the exchange

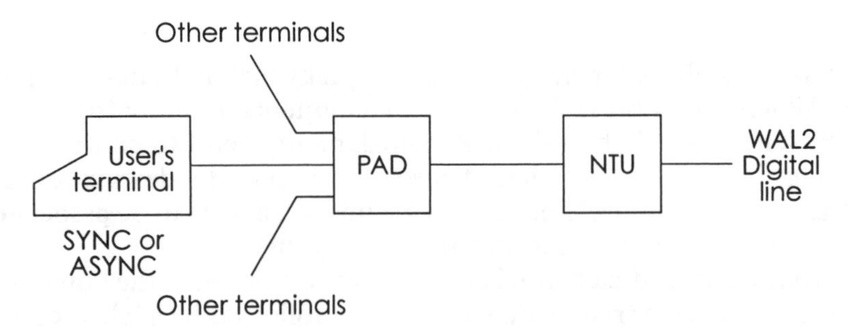

Figure 6.16 Connecting a PAD

and as the packets are formed they are steered directly into the packet switched stream. Figure 6.17 illustrates this. However they are used, PADs contain sufficient intelligence to act on command to set up calls, keep account of usage, and determine details of operation such as those listed in CCITT standard X3, which defines the command set through which a PAD may be controlled.

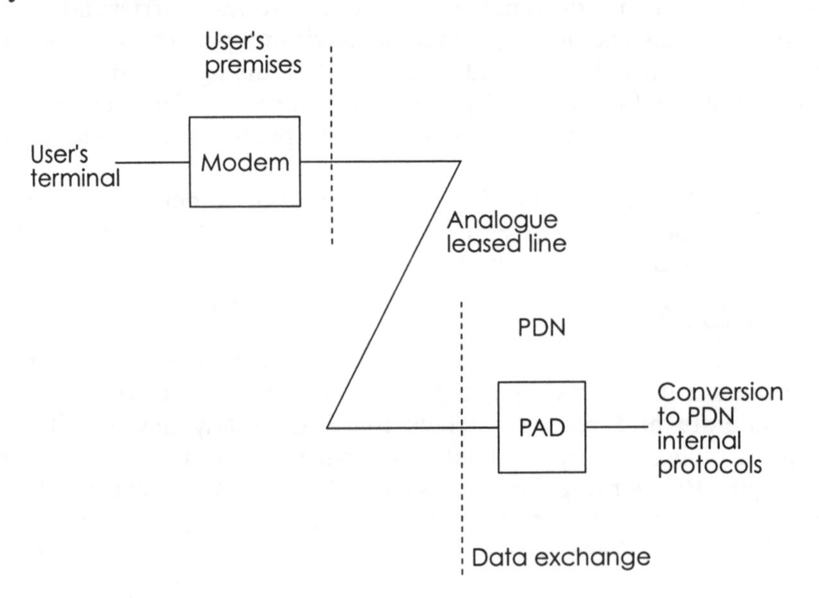

Figure 6.17 Exchange based PAD

6.3.3 X3 and the PAD

The PAD interfaces a terminal or host to the X25 system and in doing so it provides some services to the user. A PAD may be put into 'command mode'. Once this is done, any input from the terminal or host will communicate with the PAD itself and will not be passed through to the X25 system. Switching is usually achieved by a specially agreed character or symbol. Standard X3 defines a set of parameters under which the PAD operates. Command mode allows these parameters to be defined or changed. A PAD needs to be 'initialised' when first installed and this initial configuration will determine the default values that will apply whenever the PAD is switched on. Table 6.2 lists the first twelve parameters and their meaning and possible values.

Number	Meaning	Allowed values
1	escape from data transfer state	on/off
2	echo	on/off
3	data forwarding characters	various combinations
4	data forwarding timeout	0–12.8 seconds
5	flow control by PAD	on/off
6	PAD service signals suppressed	on/off
7	PAD action on 'BREAK'	none/interrupt reset/break command mode
8	data delivery to terminal	deliver/discard
9	padding after CR (print head delay)	0–7 character times
10	line folding (long lines)	1–255 characters
11	terminal speed	110 bps–64 kbps
12	XON/XOFF flow control	on/off

Details of a few of these command states are detailed to provide a flavour of the control available.

Echo

The PAD may be set to echo characters back to the terminal or host or it may be set on the assumption that the source device is operating in simplex mode. Simplex means that the source device automatically displays or registers the characters it sends without the need for a return from the destination device.

Data forwarding timeout

Whatever the PAD has in its data buffer when further input from the user ceases must be disposed of within this timeout value. Normally this parameter can be set to any value between the limits in 50 millisecond steps.

Padding and line folding

This is only an issue when a printing device is connected to a PAD output. The device may need a little time to set up for each line (carriage return, line feed or equivalent). This parameter allows null characters to be inserted after the carriage return to give the necessary time. Also the printing device may have a line length limit of 80 or 132 characters and to avoid overprinting at the end of overlong lines, an extra carriage return, line feed can be inserted appropriately.

6.3.4 X75 – internetworking between carriers

An important requirement when two networks are being connected is to minimise the impact the interconnection causes. Internet activity should not impose penalties on intranet traffic.

To achieve the end of matching dissimilar networks a gateway is inserted between them. The presence of a single gateway between networks raises the issue of ownership. To overcome this difficulty each net may supply a half gateway connected by a communication line. This way each owner has his own half gateway. The concept is illustrated by Figure 6.18.

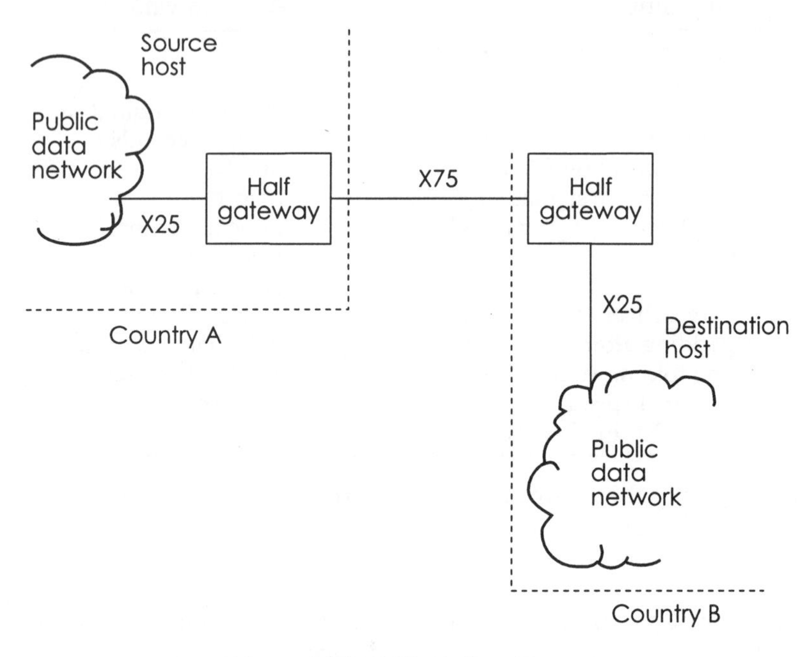

Figure 6.18 X.75 Half gateway

X75 is the CCITT recommendation defining the protocol to be followed between two half gateways, i.e. on the communication path between them.

In the model it is assumed that the source host (DCE) presents data to the digital exchange in X25 protocol. This is utilising the access protocol aspect of X25. The data, perhaps data packets, are then transmitted over the network using the internal protocols whatever they may be.

At some point the packet might be delivered to a gateway. The function of the gateway is to translate the protocol from the internal one to the internetworking standard, in this case X75. At the destination interface to another network a receiving gateway translates X75 into the internal protocol of that particular carrier's net.

X75 provides a virtual circuit operation. This means that the transport layer software module requests the network layer to set up a connection to a particular destination. The result may be that a series of virtual circuits are created.

The action described is caused by the transport layer building a call request message together with the destination address and passing this to the network layer. The network layer determines a path through a half gateway. This gateway notes the existence of the intranet virtual circuit and proceeds to set up another virtual circuit which is internetwork. The process is repeated until the destination is reached.

Messages handed down from the transport layer undergo a series of wrapping and unwrapping actions as they are passed along a chain of network level operations. It is obvious that the originator of data in one network is unlikely to understand the total topology of another and this leads to the idea that the initial gateway is chosen by the initiator and then the actual routing is devolved to the carrier networks along the way.

There is a serious disadvantage to this for sensitive traffic. If control is not exercised data could go via a potential enemy country! In this case source routing, where the route is fully specified at the origin, may be employed. Considerable knowledge of the carrier networks is inferred.

The X75 protocol is almost identical to X25. Frame and packet specifications are the same, the only difference lies in the call request packet. In this case there is an extra field before the facilities. This extra field is called the network utilities. The structure of the utilities field is as follows. The first item is a field length word allowing 0 to 63 bytes. The actual significance of the content of utilities is not laid down within the X75 specification. One requirement is documented. This is that each network involved must record its special 4 digit X121 number within the field.

6.3.5 Integrated services

The integrated services digital network (ISDN) is a system based on the concept of supporting voice and non-voice services within the same environment. The idea has been pursued for about 20 years but in the late 1980s trials took place, swiftly followed by services offered by a number of carriers. Two major options are specified, a narrowband and a broadband system. The broadband system will be available in the mid 1990s, but the narrowband option is currently available. It is the narrowband ISDN that will thus be considered. One of the main problems with the diversity of services offered is that each has its own associated jargon. ISDN is no different. A conceptual ISDN link is shown in Figure 6.19.

Connection to the ISDN is via a basic rate interface or by a primary rate interface. The basic rate type consists of three channels. Two are basic (B) type. For these, data is circuit switched into the 64 kb per second (B) links. The third channel is the basic (D) type. The (D) channel operates at 8 or 16 kb per second and is used for call set ups, control signalling or low speed data such as telemetry. The overall basic rate interface requires sufficient bandwidth to operate at 144 kb per second.

There are two kinds of access. Which is used depends upon whether the terminal equipment (TE) can generate the ISDN protocol. If so, the connection is via what is called an 'S' interface. If not, then a 'terminal adapter' is provided. The terminal adapter meets the S interface standard at its output and provides an 'R' interface to the non ISDN TE.

Connection to the digital exchange is via a 'U' interface and a network terminating box is necessary to connect the conditions of the U interface to that of the S. Figure 6.19 incorporates the ISDN terminal reference model.

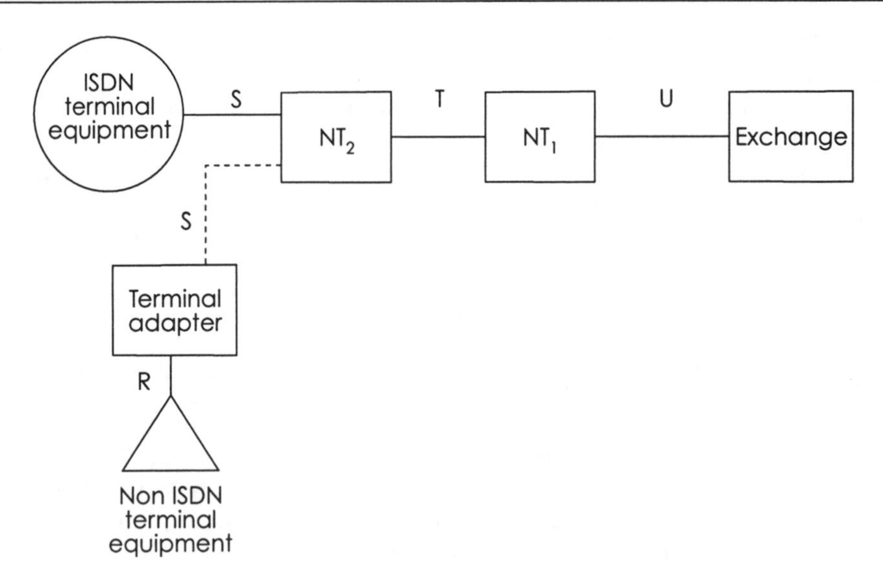

Figure 6.19 ISDN system

6.3.5.1 The U interface

The U interface uses existing copper pair technology. Because the original use of these pairs was to carry analogue speech up to 4 kHz a number of basic electrical characteristics including crosstalk from both near and far end limit the data rate. The problem can be brought into focus by remembering that for ISDN at least 144 kb per second each way is needed. The standard multipair cables used for local loops are fertile ground for crosstalk.

Many schemes to combat these difficulties have been considered but the CCITT recommend the hybrid echo cancelling option together with a special line code. The purpose of the line code is to increase the data rate without increasing the signalling rate. This notion, it will be remembered, is the basis of high rate modems for analogue circuits. With ISDN in the USA and UK the line code used is called 2B1Q. It codes two binary digits as one quaternary element. A quaternary element or Quat is one of four possible levels: + 3; + 1; – 1; – 3. The quats are transmitted at 80 kbaud. The baud is a traditional unit of signalling speed which defines the time for one signalling element. In fact the baud rate is the reciprocal of the element time. Because of the 2:1 relationship between bits and quats the system is equivalent to signalling at 160 kbit per second, which accommodates 2B+1D channel plus a little extra for control and framing purposes. Data is transmitted in a frame of duration 1.5 milliseconds. In this time 120 quats are transmitted and the first nine, (corresponding to 18 bits) carry a synchronising pattern of +3, +3, –3, –3, –3, +3, –3, +3, +3 or the inverse. The full 240 bit frame arrangement is shown in Figure 6.20. These frames are transmitted as part of an eight frame multiframe.

bit	1–18	19 – 234	235	236	237	238	239	240
		Data	M1	M2	M3	M4	M5	M6
1	ISW		Address			ACT	Control	
2	SW		d/msg			DE-ACT	& status	
3	SW	2B & D	eoc1					
4	SW	channels					C	
5	SW						R	
6	SW		eoc 2				C	
7	SW							
8	SW							

Figure 6.20 ISDN 240 bit multiframe

6.3.5.2 The S interface

The network termination unit is provided in two parts, NT1 and NT2. Each has a specific role. NT1 is used as a power extraction unit. The transmission line can suppy minimal energy for emergencies and sufficient for 'wake up' actions. The proper supply for the interfaces should be provided locally. The S interface is really the access point for users. The expectation of the S interface is that data is offered in frames at 4000 frames per second. The frames are 48 bits long and have a complex structure reflected by Figure 6.21. These frames are offered to the NT2 unit and the format involves alternating content of both B channels and the D channel. Other bits are interspersed to provide framing (F) and DC load balancing (L).

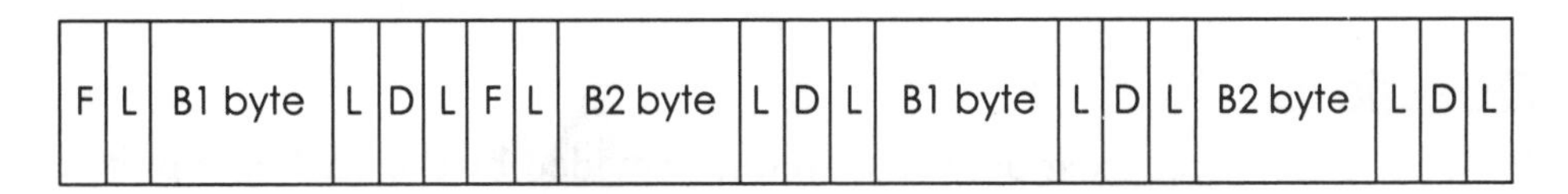

Figure 6.21 ISDN 48 bit dataframe

6.3.5.3 The contention access method

The S interface allows up to 8 terminal equipments (TEs) to contend for access using a collision detection method. The S interface frames are signalled as a ternary line signal. All signalling is subservient to timing derived from the NT unit. A TE may only use B channel blocks when they have been allocated by the NT unit. Requests for channels are carried over the D channel and the NT echoes the bits signalled back to the TE. A wired OR logical input to the NT means that any '0' will overide any simultaneously signalled '1' and so, where a contention occurs, one TE will continue unaffected whilst, by virtue of the echo and a comparison between what was sent and what was echoed, another TE will detect a 'collision' and stop sending. To facilitate comparison echos from the NT are delayed by 2 bit periods.

6.3.5.4 The data link layer

The signalling at S and U interfaces is part of a hierarchy. The bytes of the B channels and the bits of the D channels form data frames complying with the link access protocol B or D format from the X25 standard. For the D channels link access protocol D (LAPD) is used. LAPD differs from basic SDLC or HDLC systems by having the address field interpreted as two separate parts. Part one is the SAPI or service access point identifier. The purpose of the code in this field is to identify whether the frame is carrying telephony or data or video signals. The second part identifies the TE's own local address code. The operation of basic SDLC is covered in Chapter 3, Section 3.5.6.

6.3.5.5 The primary rate interface

The primary rate service accommodates 16 B channels, a D channel and 15 more B channels in a primary rate frame format as shown in Figure 6.22. This service requires a bandwidth suitable for a bit rate of 2.048 Mb per second (1.544 Mb per second in the USA). The general expectation is that large companies owning corporate networks will access the ISDN via the primary rate service.

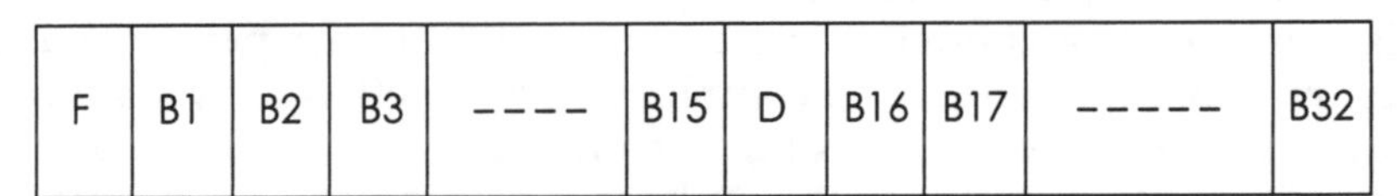

Figure 6.22 ISDN primary rate format

Questions

1 What differences exist between circuits provided by the analogue public switched telephone network and leased analogue lines?

2 Compare and contrast a link set up using the Kilostream service and one for the packet switched stream.

3 Which parameters of a sinusoid can be varied in order to carry data transmissions?

4 In the more sophisticated modems, scramblers and descramblers are employed. What is the reason for including these circuits.

5 Describe customer access to the Integrated Services Digital Network (ISDN) and explain the options in terms of channel speeds and definitions.

7

Going mobile

7.1 Cellular radio

Mobile stations provide the primary delivery mechanism for the service offered by the case study organisation. Increased flexibility and reduced response time are attainable if mobile stations can be quickly and easily directed to the next client at the breakdown site. Control and command are sometimes achieved by radio telephony alone but more modern systems could provide detailed guidance of the rescue driver without him or her needing to consult maps. Such a system requires a data link for facsimile and printed instructions. Automatic data logging of jobs and their completion would be a useful spin off.

The only system capable of national operations of this kind is the cellular radio network. The cellular system works on the idea that a network of base stations provide overlapping coverage as depicted in Figure 7.1.

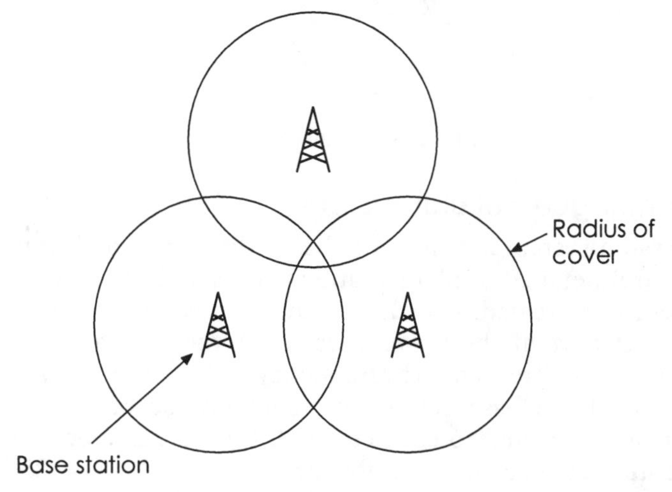

Figure 7.1 Cellular base station distribution

The size of the cellular regions rather depends on the number of users expected. In a densely populated inner city zone the cells would be quite small. There is a good logical reason for this size variation. Each base station has only a limited channel capacity and to avoid running out of capacity, the population and thus the traffic offered, is simply subdivided. The general principle of control is based on clusters of base stations, but because of the natural region shape for each base station or cell, the number of cells in a cluster may only assume certain values. The natural region shape is an imperfect hexagon and simple geometry dictates that shared boundaries can be four, seven or 12 and so on. Figure 7.2 illustrates this.

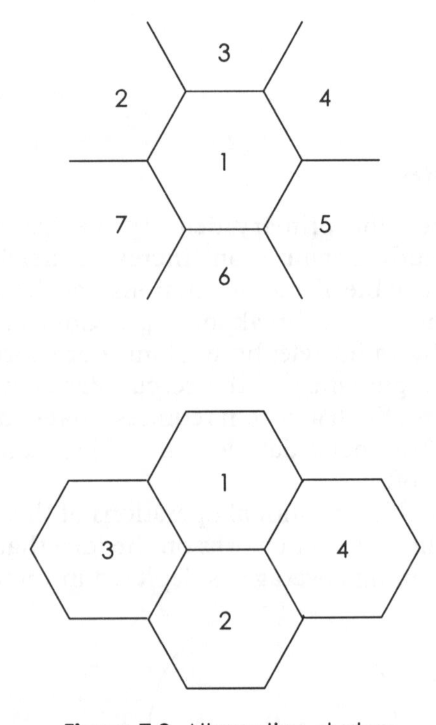

Figure 7.2 Alternative clusters

7.1.1 The analogue cellular system

The cellular radio system is constantly being developed and improved and inevitably several detailed implementations co-exist. In the early 1980s the seminal system was created. It is described as an analogue system because analogue modulation methods were used on the message channel and because the channels were created by frequency division multiplexing (FDM). The bandwidth of the base station was split on the basis of a range of frequencies being allocated to each channel. Figure 7.3 shows the channel arrangement in the spectrum. Usually channels are allocated 25 kHz each and a duplex service is normallly provided. Duplex is a term meaning communication in the two directions simultaneously and this requires two channels and thus each end, the base station and the mobile, must transmit and receive concurrently. To avoid interference between the transmitter and receiver at a given end there must be a channel separation of 45–50 mHz.

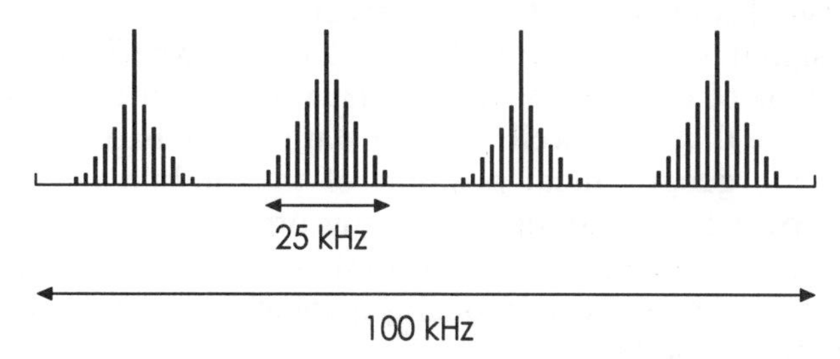

Figure 7.3 Cellular FDM spectrum

Figure 7.4 shows the UK system separation scheme. The overall bandwidth allocated is organised in two parts, with each channel having a presence in each part. The higher part is used exclusively for base station transmission whilst the lower is for mobile station transmit frequencies.

Each base station is allocated a number of channels. Normally there are two so called message channels, one from the base station to the mobile and one for the reverse direction. Control of the link is obtained through two control channels, once again one in each direction.

The irony of the analogue system is that control signals are actually digital in nature. The forward and reverse control channels are first used to negotiate and set up the message channels. It is through these control channels that a mobile station is contacted or vice versa. Signalling takes place using data frames compliant with the high level data link control protocol (HDLC). The channel carrier is deviated by 6.4 kHz in a frequency shift keyed system. The data is Manchester encoded to provide an 'in signal' synchronising element (see Chapter 2, Section 2.2.1 and Figure 2.10). The error correction system is Bose Chauduri Hocquenghem coding as described in the Chapter 4 discussion of linear block codes. Use of error correcting codes is called forward error correction (FEC) because no return path for repeat requests is

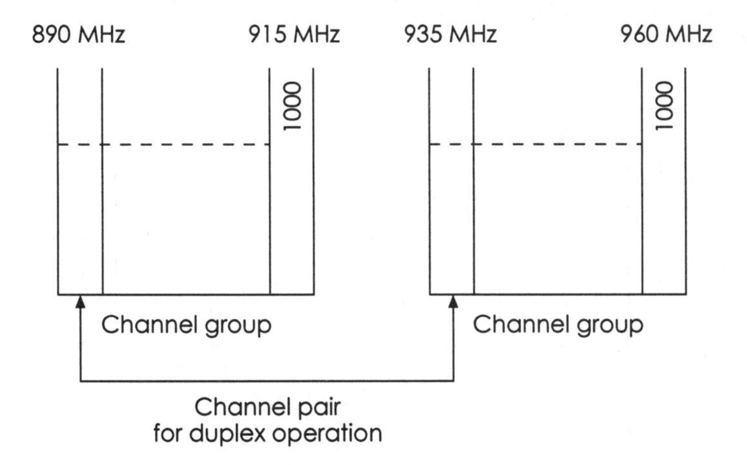

Figure 7.4 UK duplex channel separation

needed. The use of some kind of error correction is vital because of the large variations of signal strength and electrical noise experienced with radio systems. The control channel is used to set up links and checks are made regularly to ensure that all is well. Mobile stations are handed from one cell to another as the geographical position of the mobile changes and this generates handover handshaking and many other types of control signals such as registration, which is used by mobiles to announce their current position. This also allows the supporting network to route calls to the right cell for a particular mobile.

7.1.2 The digital cellular system

The analogue system had six major variants implemented in different parts of the world, all of which were and are mutually incompatible. The second generation of cellular radio systems represented a second opportunity to attain a global standard. This opportunity has again been missed, with separate systems being developed in Europe, the USA and Japan. The European system is rather hopefully dubbed GSM, the global system for mobile communications. One thing that all options have in common is the rejection of frequency division multiplexing as the underlying channel allocation scheme. The main disadvantage with FDM is the the total transmitter power must be made available in a shared way across many channels. New systems, such as GSM use time division multiplexing multiple access. In this mode the user has sole use of all the transmitter power and all of the bandwidth of the system but only for short periods of time. Other users are interleaved in time. Each user thus has a time slot during which a burst of digitally signalled data may be sent. Figure 7.5 demonstrates wide band frequency division multiplex. If the overall bandwidth were shared between a small number of channels, all subject to TDM, then the system would be operating a narrowband TDM system. There are a number of other advantages claimed. Among these are :

1. Since mobile stations transmit only intermittently in their own allocated timeslot then co channel interference is reduced because only a tiny fraction of mobiles are transmitting at any one time.

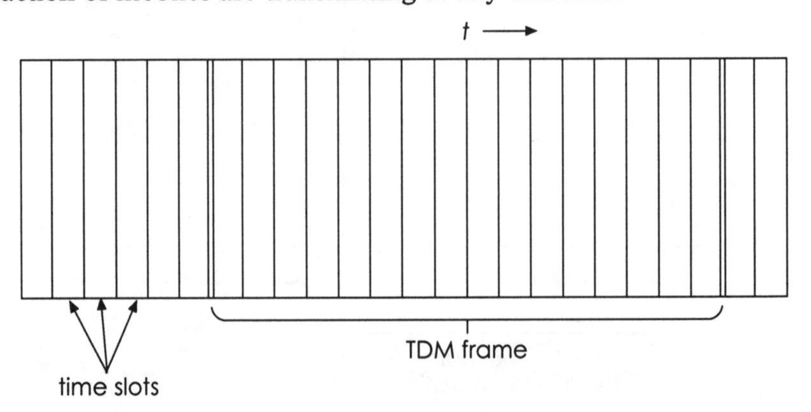

Figure 7.5 Wide band FDM

2. The number of carrier frequencies handled by the base stations is much reduced. Analogue FDM would require a carrier for every channel.
3. The filters and duplexer circuitry necessary for simultaneous transmission and reception in analogue systems are not needed. Rapid switching circuits replace these systems, allowing transmission and reception sequentially by using different slots. The sequential switching may be sufficiently fast to create the illusion of continuous connection and thus simultaneous action. The cost of switching circuits is much lower than the cost of complex filtering circuits.
4. Upgrade capability. Switching systems and digitised voice channels lend themselves easily to microprocessor control. The functionality of a dedicated microprocessor may be altered by changing the software by changing a read only memory chip. This point may be readily understood by considering improving algorithms for voice encoding. As these improve further it is likely that fewer bits per sample will be needed. Lower bit rates per channel might allow adjustment to slot lengths and protocols. Although a microprocessor could be used to control the analogue system, the analogue system cannot easily be changed because it is a function of analogue circuitry and not of digital processing.

7.1.3 GSM and the ISO 7 layer model

In Chapter 1 the ISO OSI model was introduced as a general principle for the stepwise decomposition of the problem of providing a digital network. In Chapter 6 normal line or optical fibre services of data carriers were explained and in particular reference was made to the scope of X25. In that discussion it was suggested that the carrier would not be concerned with issues above layer 3. The reasoning was that layers 1,2 and 3 provided an error free communication service and that more sophisticated functions were the responsibility of the user. The same logic can be applied to the providers of radio links. For this reason the design of GSM although based on the ISO model, specifies only layers 1 to 3. The issues covered by each layer will be rehearsed to allow a comparison with non radio systems such as X25. The GSM model relates to the ISO model as follows :

Layer 1 – Physical

Under this heading the GSM includes the radio frequencies used, the modulation type, the structure of TDM signalling, and error control coding. This last feature is unique to radio links simply because the signalling of individual bits cannot be guaranteed without error.

Layer 2 – Data link

Time slots do not correspond to data frames and so the frames must be broken down and transmitted in parts and at the receiving end the parts must be concatenated to reform the frames. Clearly, the frame protocol of acknowledgement is a feature of this level.

Layer 3 – Network layer

In a traditional environment, data packets are dealt with at this level. In a Virtual circuit packet system the format and protocol defines how the message will be delivered through the network by providing full delivery addresses in a call set up phase. There is a slight adaptation necessary to see how a radio system relates but the key point is the realisation that a virtual call set up is a communication management function. GSM requires a call management function, albeit more complex because not only call control management but also mobility management when mobiles move from one cell to another and radio resource management because the system is a TDM based system. Figure 7.6 attempts to relate the functions and entities in a simplified way.

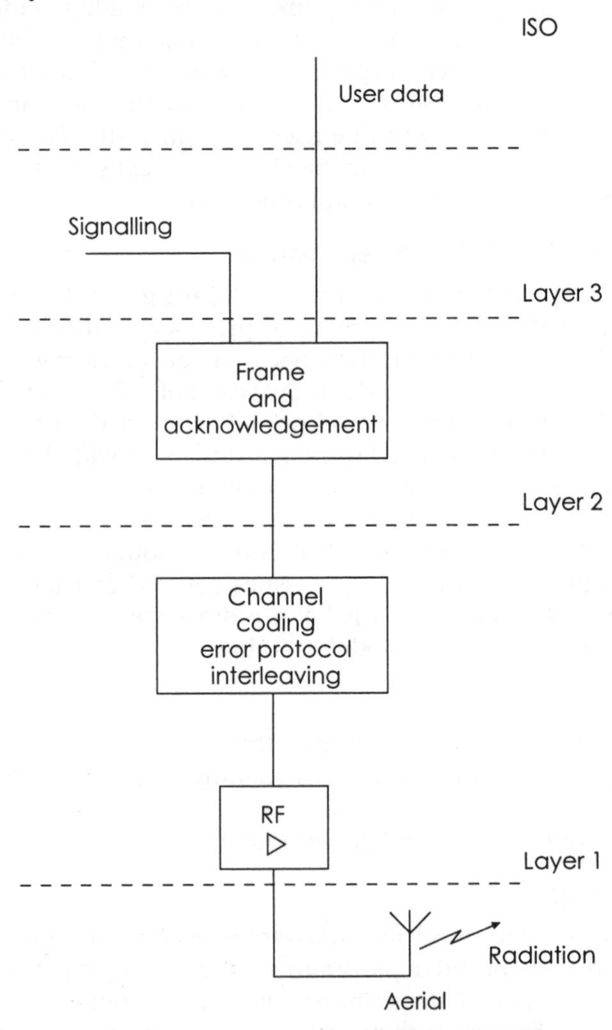

Figure 7.6 Layer model for GSM

Table 7.1 compares the performance of analogue and digital systems.

System	Analogue (TACS)	Digital (GSM)
Multiplexing	Frequency division	Time Division
Channels per carrier	1	8
Carrier spacing	25 kHz	200 kHz
Number of carriers	400	50
Number of channels	400	400
Data rate	10 kbit/second	270 kbit/second

7.1.4 GSM infrastructure

Each base station in the cellular structure must be connected to both the public switched telephone system and to the integrated services digital network (PSTN and ISDN). This requires a switching centre or exchange. In cases where clusters of base stations are in dense profusion one switching centre can serve them all. The switching centre is computer controlled and uses the PSTN or ISDN to pass information about mobile station locations. Each mobile station possesses an identity that is registered at a particular switching centre. This switching centre is the home base of the mobile station. When a mobile station visits a 'foreign' area, the foreign base station uses the fixed links (ISDN/PSTN) to refer back to the mobile's base. This can be done because part of the call set up procedure involves the mobile in providing a coded number to the communicating base station. The exchange of information between the foreign and home base stations allows for authentication and details of any supplementary features allocated such as special data services. The refer back process is initiated when an active (switched on) mobile moves to a foreign cell area and by this means the home base will know the location of a roaming mobile. Calls for the mobile may thus be routed correctly. A major control feature of cellular radio systems is the handing over of mobiles transiting between cell areas. GSM uses the time slot system to insert 'broadcast control channel' data which includes base station identity, frequency data and frame synchronising signals. A mobile monitors up to six other cell control channel transmissions and reports signal strength and other valid parameters to its current base station. In this way, candidate hand over bases are tracked and when a candidate clearly produces a better channel a hand over is initiated.

The full details of the operation of cellular systems are outside the scope of this text because the service is primarily aimed at providing speech channels. GSM processes can be related to the ISO 7 layer subnet as shown in Figure 7.6. GSM does expect users to exchange data transmissions and allows data rates of up to 9600 bits per second if the user supplies necessary data protection measures or 4800 bits per second as an error free link. Inevitably, the increased use of ISDN services on the fixed networks will result in offerings over the cellular radio system in due course. The point about either the analogue or digital cellular system is that users may send data digitally over them using phase shift keying similar to the methods used in the modem technologies covered in Chapter 6. The limitations and constraints relate to the fact that both options are optimised for telephone voice operation. Data specific cellular radio systems have been devised. An example packet network is PAKNET.

7.1.5 Packet radio network (PAKNET)

The fixed telecommunications system offers an access protocol for the public digital data network. The access protocol is the familiar X25. X25 implements the ISO model to layer 3 and is a packet system. A cellular radio structure for packet transmissions through mobile connection to base stations has been devised based upon an allocation of frequencies in the VHF band (very high frequency band – 30–300 MHz). This range is well below the bands associated with analogue and digital cellular which is around 800–950 MHz. The benefit of a lower frequency for the channels is extended range, perhaps two and a half to three times greater in this case, since PAKNET operates in the region of 160 MHz. For this reason PAKNET covers larger cells than the standard system. Table 7.2 shows the frequencies associated with PAKNET base stations. It is notable that each base station has only 14 channels available. A further 14 frequencies are allocated to mobiles, yielding seven channel pairs. A channel pair is necessary for each link. The access protocol is multiple access and is of a dynamic slotted reservation type. The data rate is 8 kbit/second. The base station seperates time into access or contention periods and data transfer periods. Several 27 millisecond slots are allocated in each period and any mobile station wishing to send a data packet waits for the base station to signal the contention period and then waits further until a specific timeslot when the mobile transmits. A transmission in a given timeslot reserves an associated slot in the data period. The mobile, having made a reservation, waits patiently until the base station signals the start of the data transfer period. At the beginning of the appropriate timeslot the mobile issues its data packet. No collisions actually occur because the perfect feedback effects of radio systems mean that all the other stations know that the slot is reserved. If two or more mobiles try to reserve the same slot then a recovery strategy must be applied. Typically this entails throwing away a contested slot, but there are other possibilities which will be explored later. Figure 7.7 illustrates the events. This arrangement is much simpler than for ordinary cellular communications because there are no discrete control and message channels. A single channel is used for control and messaging.

Channel	Base transmit (MHz)	Base receive (MHz)
2	164.2250	159.7250
3	164.2375	159.7375
	and so on until	
15	164.3875	159.8875

The performance of a time slotted contention system of this kind may be arrived at by establishing the equilibrium state occupancies of the slots, together with the mean throughput and delay. The PAKNET is configured to accommodate 800 mobiles or fixed outstations per base station. The built in assumption for this population is that the average rate at which stations generate packets is one every 15 minutes. The packet size is 128 bytes of user data signalled as a 12,12 code. Chapter 4, Section 4.4 provides an equation for calculating the error correcting capacity of such a code.

7.2 Satellite systems

Satellite systems are increasingly being used for communications and for position fixing. The case study might find considerable use for a system which enabled rescue units to be pinpointed in remote rural and perhaps mountainous landscapes. A satellite communication system (Satcom) has characteristics which were inconsequential in earlier times but which have become significant because of the changing demands of users.

When satellites were first launched users were used to the circuit switching paradigm and they were also happy with the notion that all communications would be sent via the public carrier or PTT. Public carriers used satellite channels in the same way that they would use the channels available on an undersea cable. The greater sophistication in user requirements such as packet switching have meant that users need direct access to the satellite. Whereas the old notion allocated all of the bandwidth of a channel the new ideas require the channel to be shared. The one characteristic that becomes highly significant under a regime where a channel is shared and direct access is permitted is propagation delay. To remain in geosynchronous orbit a satellite must be positioned 22 282 miles above the Earth. Figure 7.7 shows the effect on the distances between earth stations. The satellite acts as a re-broadcaster or relay. The altitude of the satellite will create a minimum two way travel time of 240 ms, even when the satellite is directly overhead for

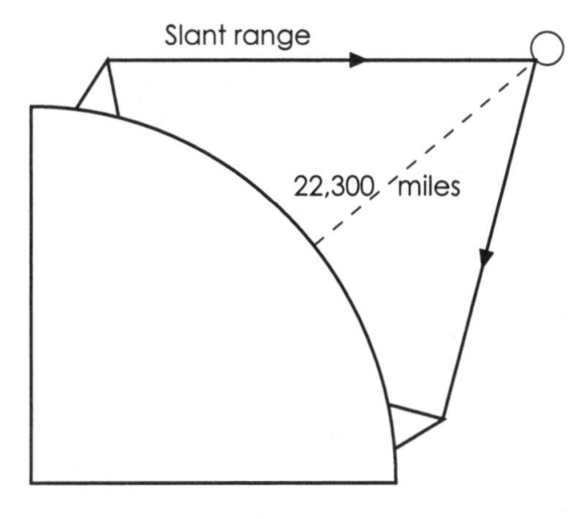

Figure 7.7 Slant range

sender and receiver. The more realistic situation involves a slant range and a longer delay. There are implications for the access method used by groundstations.

7.2.1 Polling

Shared channel access may be arbitrated in several ways, one of which is polling. In the context of satellite systems the Master station could be on the ground or in the satellite. If the satellite acts as master then it polls the groundstations one at a time. A single poll will take 120 ms minimum to reach the intended groundstation and another 120 ms minimum for the station's response to reach the master. 240 ms for the round trip means that if there are 200 groundstations, then if a particular groundstation response is negative, that station will have to wait 48 seconds before it is polled again. Such delays would be unacceptable. Figures 7.8(a) and 7.8(b) illustrate the point. In Figure 7.8(b) the master is assumed to be on the ground and the delay between polls will be 1 minute 36 seconds. Polling does not fit very comfortably in a system that has a dynamic population of nodes or groundstations and for this reason too it is rather problematic.

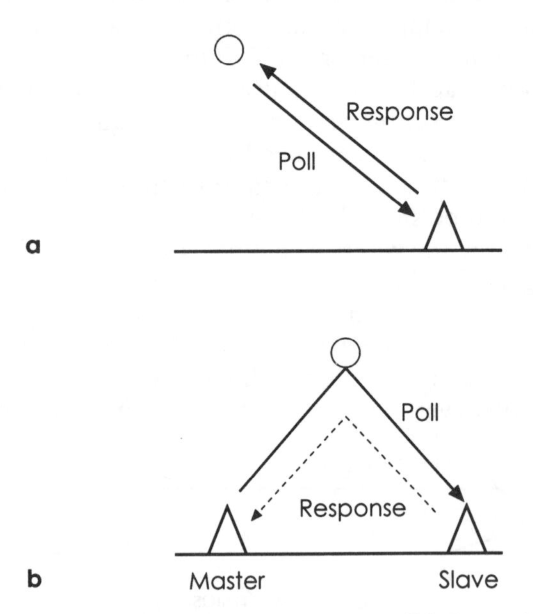

Figure 7.8(a) Satellite polling access, **(b)** Ground polling access

7.2.2 Token access

Again token access proves wholly unacceptable. Once more it is the propagation delays that prove fatal. Each 'token pass' will require the token to be sent up to the satellite and be returned down to the next in line. The two way travel time is 240 ms. A population of only 100 stations will have possible token possession only every 24 seconds minimum, and much more if stations have data to send.

7.2.3 Time division multiplexing

Figure 7.9 shows how time is apportioned in a typical time division multiplexing access (TDMA) satellite system. As with terrestrial time division systems time is split up into a series of slots. Preceding channel information there is a supervisory preamble burst.

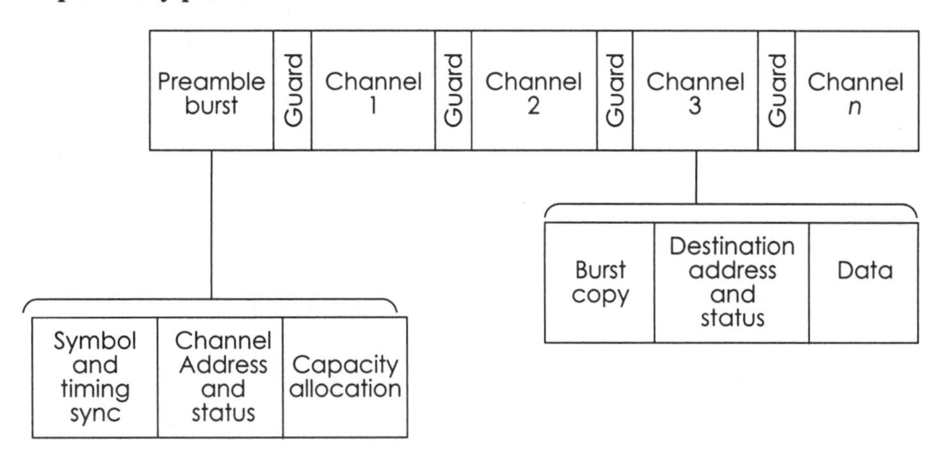

Figure 7.9 Satellite TDMA

The preamble burst serves to set up symbol and burst timing for the following TDM frame. TDM frames run continuously but do have a 'length', which means that a maximum number of stations can be served. The number corresponds with the number of slots in the TDM frame. This correspondence is a limitation and the length of the frame must be well judged to avoid unneccessary delay whilst stations wait for their time slot to come round. Each time slot is 'owned' by a particular station and this leads to possible inefficient use when some stations are offering very little traffic as their time slots go idle. The guard bands in between channel slots in the frame are short time periods that ensure the separation of channels and the avoidance of overlaps in transmission. Figure 7.10 is a typical TDMA system. The uplink is shown as 14 GHz whilst the downlink is 12 GHz. The choice of frequencies in that part of the electromagnetic spectrum enables physically quite small aerials to be used. This is a strong advantage in mobile applications.

7.2.4 Contention

Apart from TDMA, with its inherent potential inefficiency and inflexibility, there is only one sensible access method. Contention could be used in its simplest form, with stations simply transmitting data to the satellite as and when required. The whole bandwidth of the satellite system could be used as a broadcast network equivalent. The perfect feedback available because all stations can listen simultaneously enables data collisions to be detected fairly easily. There would be a large number of collisions in this application when compared with a cable system because of the large window when collisions

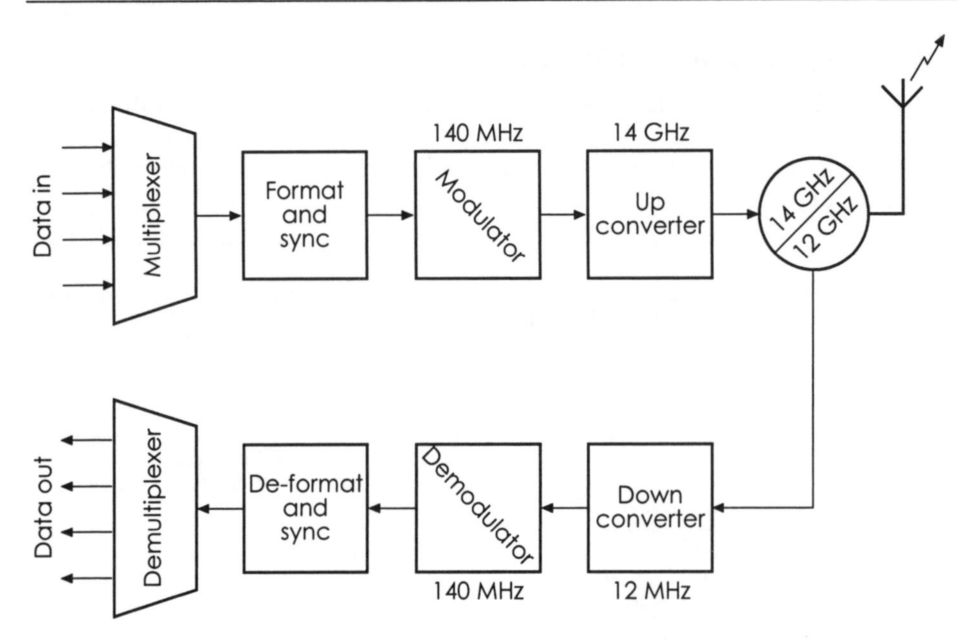

Figure 7.10 A typical TDMA system

may take place. The window size is a direct function of the worst case propagation delay produced by the broadcast highway. Line based networks will have delays of the order of a few microseconds but satellite systems, because of the very large distances involved, will operate with delays of not less than 240 to 270 milliseconds. The potential for collision is much greater and thus the throughput for a given amount of traffic offered will be lower. Each collision causes a delay of 270 milliseconds because it takes that long for detection. Obviously delays to an individual data packet might be much

Figure 7.11 Traffic/throughput characteristic

Figure 7.12 Delay characteristic

longer than this if multiple collisions befall it. An interesting feature of contention in this context is the relationship between data throughput and the delay of delivery of data. Assume that there is no traffic in the first instance. If some traffic is offered the throughput will rise. If more traffic is offered, some collisions will occur and although the throughput may continue to rise it will rise at a lower rate than originally. Eventually, increasing traffic offered may result in a reduction of throughput when a lot of collisions and repetitions are made. The general characteristic is shown in Figure 7.11. If the delay in delivery is tracked during the same period the result is as illustrated by Figure 7.12.

7.2.5 Stability

If a simple contention system is used when long propagation delays allow large contention windows then there is a possibility, if the traffic offered becomes very high, that throughput will fall to zero. Another way of thinking of this state is to imagine that there is a gridlocked condition. Such a state is defined as 'instability' and the problem is that once the system enters instability it cannot recover. Gridlock is permanent. It is possible to achieve a non-negligible throughput when throughput is backlogged by using a slotted contention scheme but generally the price to be paid for unconditional stability is very long delays in delivery. The big advantage of time division multiplexing is that instability is impossible. During the development of satellite based data systems the good features of contention, namely, low delay at low traffic levels and no limitation in numbers of user because of fixed slot allocations, were considered desirable. At the same time, the good feature of time division multiplexing, namely, unconditional stability was also considered desirable. A number of schemes were devised which attempted to gain advantage from both techniques. These combinational ideas are called 'reservation' systems. All of them operate as a slotted contention system at low traffic levels and gradually introduce an element of slot reservation or ownership as traffic offered increases. There is a gradual transition to time division multiplexing. Where the methods suggested differ is in the way slots are reserved and subsequently how they are released should traffic fall.

Questions

1 Interconnection of company LANs over long distances may be achieved by radio, including satellite systems. With particular reference to access methods, discuss the constraints that render all but contention and time division multiplexing invalid. Comment on the proposition that a dual access method, where a system normally operates a contention system but falls back to time division multiplexing when throughput falls, has advantages.

2 Explain the main differences between the analogue cellular system (TACS) and the digital cellular system (GSM).

3 Bearing in mind that both TACS and GSM are optimised for speech transmission, discuss the simplest way in which data signals could be carried.

4 Outline what usage or geographical consideration would prescribe a radio network distribution system.

5 It is sometimes claimed that 'flooding' is an appropriate routing method in radio nets. Seek to justify this statement.

8

The company network

8.1 Introduction

In Chapter 1 the case study was introduced and three stages for the implementation of the company network were suggested. At that point only stage one, the head office network, was specified in detail. After a comprehensive review of the techniques and technology appropriate to local area, medium range private and long haul working options provided by public carriers it is now appropriate to integrate them into an overall company network. The original requirement specification was for:

- internal networking in company buildings
- relatively local interconnection with satellite buildings
- longer range interconnection to branch offices in major cities
- radio interconnection to mobile units in metropolitan and normally rural areas
- radio interconnection to mobile units in remote highly rural areas.

It has become apparent that a modern organisation would require to handle data of all kinds including: alphanumeric data; line drawings with textual annotation; photographic images; and speech. In the context of the case study, a national car breakdown and recovery service, each of these forms of data is relevent. The regional office despatcher may wish to communicate 'next job' instructions to mobiles as a written order supplemented with a road map showing the location of the job. All activities would have to be recorded for statistical and accounting purposes. A centralised data base held on a head office server would have to be regularly accessed with alphanumeric data records. Photographic image data may well be used to verify the identity of drivers covered by the scheme.

8.2 Head office and satellite buildings

Chapter 3 concluded that within the head office and satellite buildings the most appropriate local network would be 10base5 contention bus operating at 10 Mb/s. Given that this conclusion would also be valid for satellite buildings, the implementation of a solution to the interconnection problem is needed. Both FDDI and ATM were considered for branch office interconnection in Chapter 5. In that case the suggestion was that FDDI was the more mature system and that a range of 62 miles or so was possible. Clearly, such a range is only possible where the FDDI network is owned and operated by a telecommunications carrier. It is inappropriate for a motor rescue organisation to operate and maintain a link of this kind. Telecommunications carriers are offering FDDI links but of course the bandwidth would have to be shared with other users and there would be significant line rental charges. Where a company such as that in the case study has reasonably adjacent satellite buildings, it is possible to have an FDDI link wholly privately owned and operated by the company. Arrangements would have to be made with the local highways and footpath authority to lay the cable (optical fibre) trunking and the installation. Figure 5.9 shows a layout for networking the head office and satellite buildings. Note that the FDDI backbone interconnects the buildings and that the FDDI double ring is connected to the 10base5 local nets via FDDI–Ethernet bridges. In Figure 5.9 one building has a token ring installation. In the case study this would be replaced by another FDDI–Ethernet arrangement.

8.3 Head office to local regional offices

In Figure 5.9 the head office 10base5 network has two FDDI bridges. An additional bridge could interface to the telecommunications carrier's long distance FDDI ring. At the local regional offices an identical bridge connects the internal 10base5 system to the FDDI ring. The advantage of this arrangement is that the backbone provides a transparent high speed link between the offices. If telecommunications carriers are not able to offer FDDI connection then local regional offices will have to be linked by lower speed services such as those used for greater distances.

8.4 Linking to remote offices

The choice of implementation depends on the expected traffic density on the links between offices and whether a digital data exchange is available. The costs of services offered are tied by telecommunications providers to a marketing policy and tariff. The options are:

1. the integrated services digital network (ISDN)
2. the packet switched stream (X25)
3. leased digital virtual circuits (Kilostream or Megastream)
4. The analogue telephone system
   (i) leased lines
   (ii) dial up services.

If either leased lines over the analogue telephone system or leased virtual circuits over the digital data system are chosen, the provider is simply making available the links and does not provide any structure for orderly or organised data transfer. This means that no switching of data occurs. It is up to the user of the collection of links to make that collection into a company network. Consider the linking of a number of remote regional offices. The links will form an irregular network as described in Chapter 5 and illustrated by Figure 5.16. To provide the necessary routing the user must provide a switch or exchange to interconnect incoming and outgoing paths. The nature of the switch is dependent on whether it is intended to operate a circuit switched or a packet switched system. In the past circuit switching was the norm. This required the ownership of a data private automatic contention exchange (PACX) located at each of the centres formed by the convergence of leased lines. Any incoming line could be connected to any outgoing line but a single channel occupied the full bandwidth of the lines it was using. The exchange was a 'contention' exchange because the crossbar switching array at its core would typically be split into three parts. Statistically not all of the input lines would demand connection at the same time, which allows a large saving in crosspoint switching. To provide full linkage for 100 input and 100 output lines requires 100 × 100 = 10 000 crosspoint switches. If only 10 per cent of the lines are ever expected to be active the same functionality can be achieved with an input group of 10, 10 in and 1 out arrays followed by a 10 × 10 array and finally a 1 in and 10 out array. These three arrays are called the concentrator, distributor and expander respectively. The total number of switches in the three part arrangement is 100 + 100 + 100 = 300. If such a system is near capacity, an incoming line may demand a connection but find that there are no free ways through the distributor. This is the 'contention' referred to in the system name.

An alternative architecture for small exchanges uses the time division multiplexing principle. Here, it is necessary for the data to be in the form of discrete bytes, perhaps textual characters. The contention principle is evident in the statistical multiplexer. A 'straight' multiplexer divides time into time slots and simply switches incoming lines into those timeslots. At the destination a similar switch, running in synchronism with the first, demultiplexes the data. The concept is shown in Figure 8.1. The disadvantages of this setup include the loss of efficiency when a particular incoming line presents no data during its allocated timeslot and the relative lack of flexibility in connections. The statistical multiplexer does not have possibly idle timeslots. Instead, it

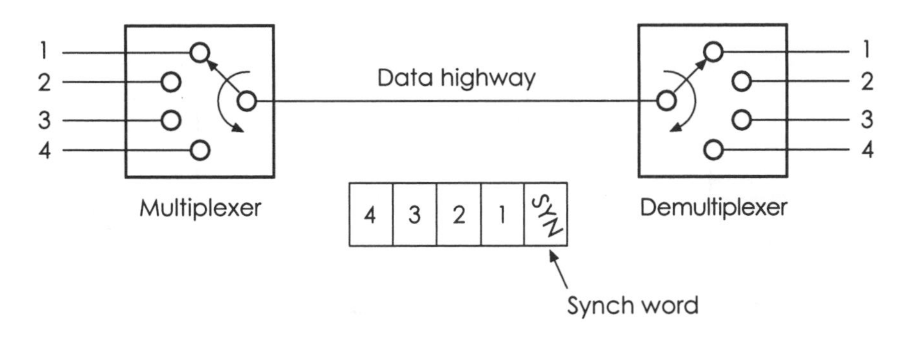

Figure 8.1 A straight multiplexer

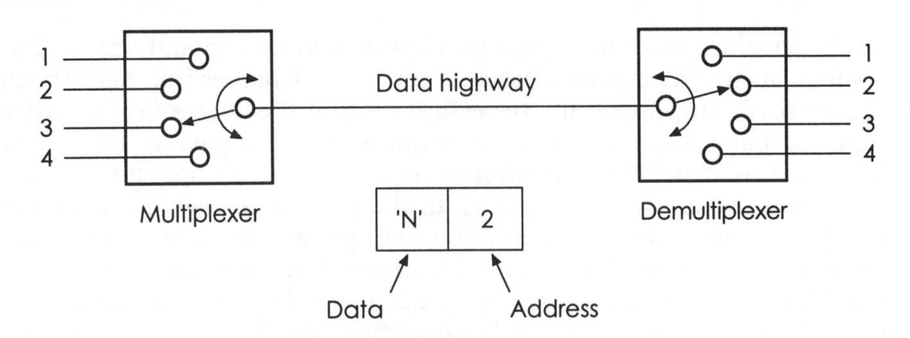

Figure 8.2 (a) The statistical multiplexer

waits for an incoming line to generate data, collects it and tags it with a destination line number. In this way any two lines may connected in any time order. Figure 8.2(a) illustrates this. In a straight multiplexer the common highway must operate at a fixed multiple of the input line data rate. If each line were operating at 1200 bits per second and there were four of them, the common highway must run at 9600 bits per second. The speed is 9600 rather than 4800 because of the need for a synchronising character at the beginning of each timeslot cycle.

On the face of it, the common highway need only run at 6000 bits per second but that value is not a standard for serial communications and 9600 is the next standard value. A statistical multiplexer may take advantage of the statistical characteristics of traffic presented and because each character requires an additional 'address' character, then the common highway data rate will only be half the signalling rate. If is assumed that incoming lines are active for 25 per cent of the time, twice as many incoming lines may be dealt with at 1200 bits per second whilst retaining a common highway rate of 9600.

How then, may a statistical multiplexer play the part of an exchange? The common data highway is not important and may be very short and fully enclosed within the system box or enclosure. What is important in this context is the ability to switch a character from any input to any output. Once

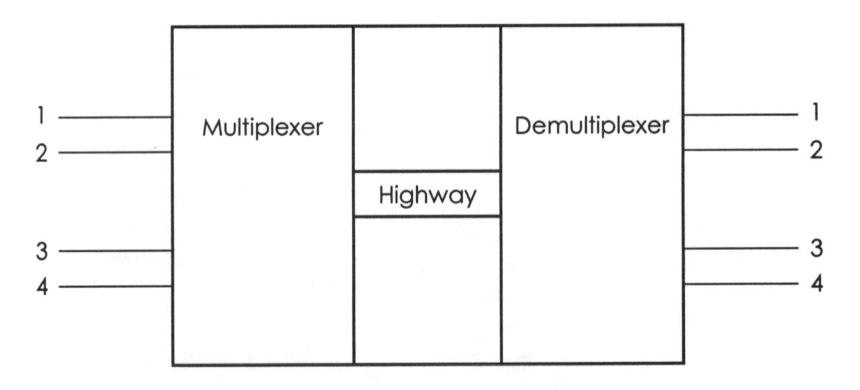

Figure 8.2 (b) The multiplexer as an exchange

a set up is defined characters will be passed through the multiplexer in such a way that the actual multiplexing is transparent. The system appears as in Figure 8.2(b). Of course multiplexers are mainly used to concentrate a number of channels down a single highway where it is uneconomic or inconvenient to lay many lines. The use of a multiplexer as an exchange is an interesting piece of lateral thinking.

Circuit switching has a number of problems when used for computer networking. Using circuit switching entails a complete end to end pathway set up before data can be exchanged. To get an idea of the delays that could accrue it is only necessary to recall the connection delays that can be evident when making a telephone call. During the delay the telephone system is forward routing and attempting to find a way to the ultimate destination. Taking the analogy a little further, a telephone connection is usually confirmed by the called subscriber saying 'hello'. For a computer link the call request signalling must reach the destination and a call acknowledgment returned. All this takes time.

The alternative is packet switching. If a user is content to use the digital packet switching service provided by a telecommunications carrier, then all that is necessary is to present data to the carrier in the prescribed standard form, using the prescribed protocol. The world standard protocol for this is X25. A company may rent or own the necessary packet assembler disassemblers (PADS) and network terminating units (NTU) but telecommunications carriers charge for every packet handled. Although the per packet charge is small, if a lot of traffic is envisaged then network costs could be high. There is no reason why a company should not rent digital leased lines of the Kilostream or Megastream type and then invest in ownership of packet switching nodes to connect them up. Packet switching is much better suited to computing because of the 'bursty' nature of data transfers. The inter-data gaps would be wasted time and capacity in a circuit switching system. In packet operations there is a maximum packet size and capacity may be used by diverse packets from different sources and for different destinations, merely coming together on a particular highway until they diverge at the next switching node. In these cases time division multiplexing is totally overt and is the basis of operations. Whereas in circuit switching, dynamic congestion is not possible (if you are connected you are connected until you are finished), in packet switching nodes short term overloads can occur. The design of a packet switching exchange or node must include strategies for buffering and queueing. The situation is described by Figure 8.3.

The packet switching node may take the form of a dedicated microcomputer with sufficient main memory to create and operate the necessary number and size of incoming and outgoing queues. Short term overload would be coped with as the queues act as first in, first out buffers. Long term congestion perhaps due to link failure or fault might be dealt with by queueing incoming data on a Winchester disk and not accepting too many more incoming packets. It may be possible in some cases to re-route packets for their onward transmission. The queue data structures and their pointer control may be run using a multitasking operating system. If this is so, then

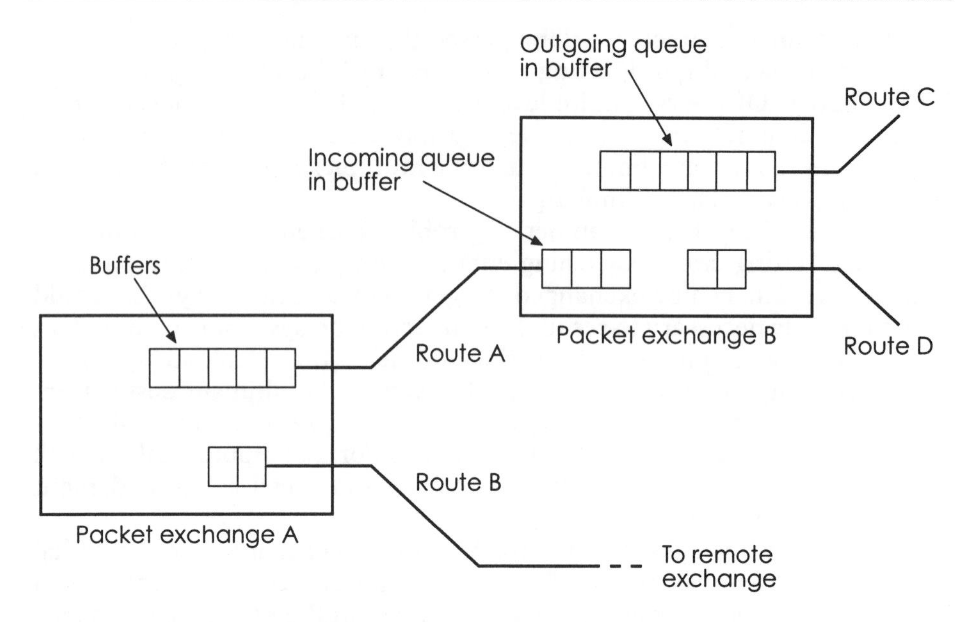

Figure 8.3 Packet switching exchanges

each queue process is dealt with as part of the round robin cycle and the resulting pseudo concurrency will mean that packet transmission and reception will be overlapped and simultaneous. Figure 8.4 gives the timing in such a scheme.

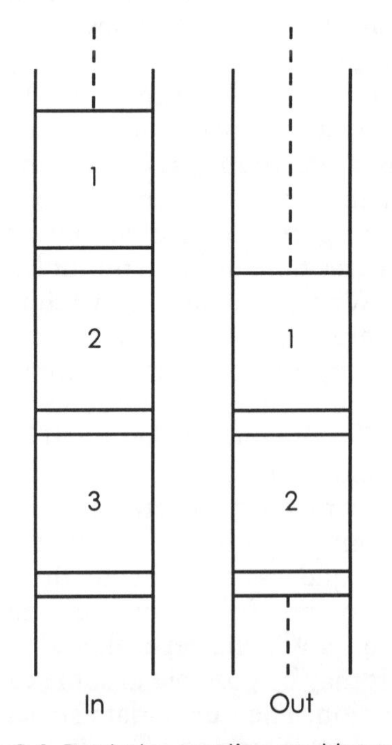

Figure 8.4 Packet reception and transmission

Packet switching was chosen for the case study and furthermore it was decided that a virtual call protocol was acceptable. X25 is specified for consistency and standardisation. It is a general rule in top down design that it is best to avoid choosing technology that is not well proven. X25 is a mature technology and would be used for links where the traffic offered was very occasional or of low density because it would be feasible and economic to use the carrier's services. If the majority of the network hardware is owned but some is provided by carriers it makes sense to use a consistent protocol . The main reason for this is that network management and technician staff will only have a single learning curve to cover most activities.

8.4.1 Routing and switching centres

Routing strategies were examined in Chapter 5. In the context of the case study, fixed directory routing is the simplest to implement. Any alterations to the network do require that network management staff update each of the switching centres. If a network is to be developed systematically with relatively little *ad hoc* addition or subtraction this should not be a problem. The benefit of a simple solution is that there are no unexpected or unpredictable results. In the early life of a network it is a sound design strategy to ensure a high level of credibility on the part of users. It takes a long time to build confidence and credibility but it only requires a few failures to destroy that credibility.

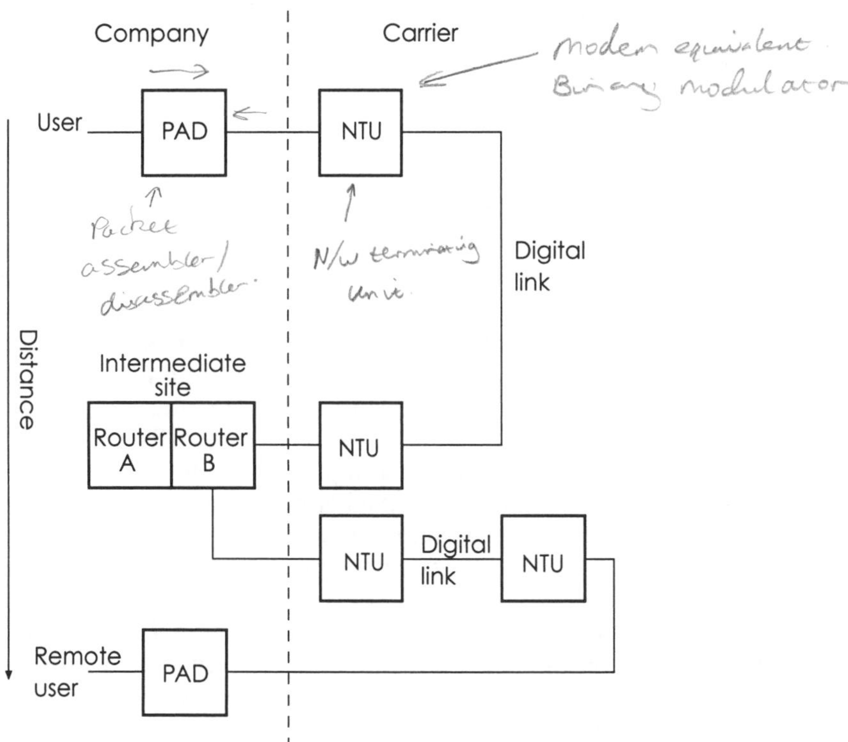

Figure 8.5 Ownership of network elements

8.4.2 Private X25 use

Assuming that Kilostream or Megastream links are in place and that X25 packet switching nodes are in place at the network switching centres there is no reason why an organisation may not use the X25 protocol as a purely private arrangement. The fact that it is the access protocol to the public data network is incidental. X25 is a virtual call, packet based system for which hardware such as packet assembler disassemblers are readily available. A private network is likely to be simple compared with the public system although it will still be necessary for the data network operations and management staff to define routing and queueing structures, as well as run and maintain the switching centres. Figure 8.5 shows the possible ownership of network elements for a private system.

In Figure 8.5, the switching centre is a router as described in Chapter 5, Sections 5.4 and 5.7 and the arrangement shown is for communication between single correspondents. The packet assembler disassemblers (PADs) shown would be termed 'monopads'. Monopads would be appropriate for connection to a gateway connecting a local area network such as 10base5 to an outgoing long distance path as illustrated by Figure 8.6. In the company

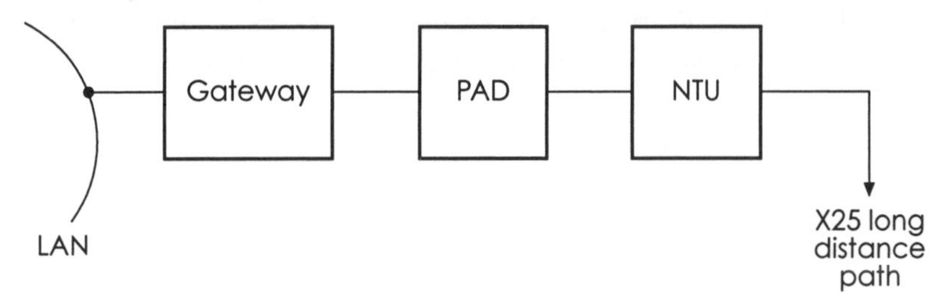

Figure 8.6 Monopad outgoing paths

network that satisfies the case study there are likely to be remote destinations that do not have local area networking. An example might be a small regional office which consists of a single room with half a dozen desktop positions, each of which needs access to the company network and to databases held at headquarters and other larger regional offices. The single room branch would be well served by a multi way PAD. One of the inherent features of X25 is its ability to handle many simultaneous virtual calls. A single PAD output through an NTU is effectively multiplexed. As described in Section 5.4.2 the format of the X25 call request packet provides for a 12 bit channel number.

In theory, up to 4096 channels could be accommodated. In practice a multiway PAD might accept call requests from a number of its inputs and concurrently set up and run the necessary virtual calls. The suggested situation is illustrated by Figure 8.7, in which a PAD has a fan in of four synchronous serial lines. Each line might be associated with a given desktop terminal or computer. PADs are operated by a command language and the command 'CALL destination_name' will result in an outgoing call request packet. Simultaneous or near simultaneous requests will result in call request packets following each other sequentially down the single line through the

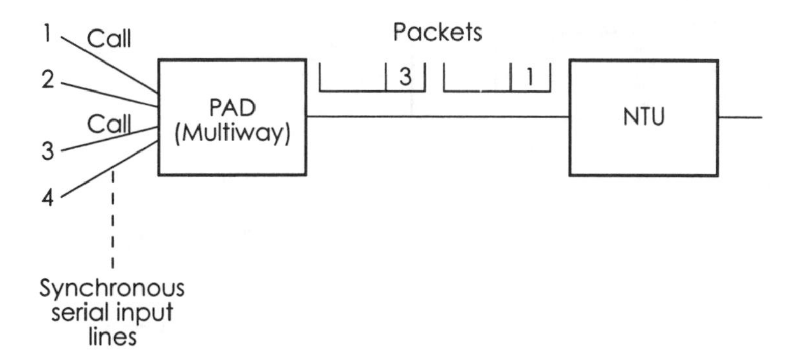

Figure 8.7 PAD virtual calls

NTU. Once virtual calls are in progress the action continues in the same way but the sequential packets are the interspersed data packets associated with a number of calls.

If the PADs in Figure 8.5 had been multi way there would be no change to the link and router architecture. At the switching centre (router), the incoming packets are sequentially received and may be switched to different output highways.

8.4.3 Public X25 use

The ownership of the full means of communication, excepting the links, but including the switching centres and PADs was justified on the grounds that high traffic density would incur heavy charges over a purely public carrier. This is because most carriers charge for packet operations on a connection charge plus a charge per packet sent. An annual charge for a digital circuit is made irrespective of traffic carried or protocol. The greater flexibility offered would have allowed the user to use a datagram format if so desired. For reasons of consistency, it was decided to use a private X25 set up. There are nearly always network nodes for which the arithmetic in terms of expected traffic and charges do not justify the expense of the private option. In cases like this the logical step would be to use a carrier's packet network. The only difference in the architecture would be that the digital line from the NTU terminates at the carrier's packet switched exchange. The packets are disassembled and passed through the carrier's network using internal protocols before being forwarded in a new X25 operation to the required destination. Figure 8.8 illustrates this.

8.4.4 Analogue links

The public digital network is growing fast but there are still some rural destinations that are not yet able to connect directly. The analogue telephone system had the advantage of having been established and operated over a long period of time and it has become ubiquitous. Because the public switched telephone network (PSTN) is based on circuit switching principles the charging regime is such that the cost of a call is simply dependent on connection time. There is also usually a small line rental charge. For very

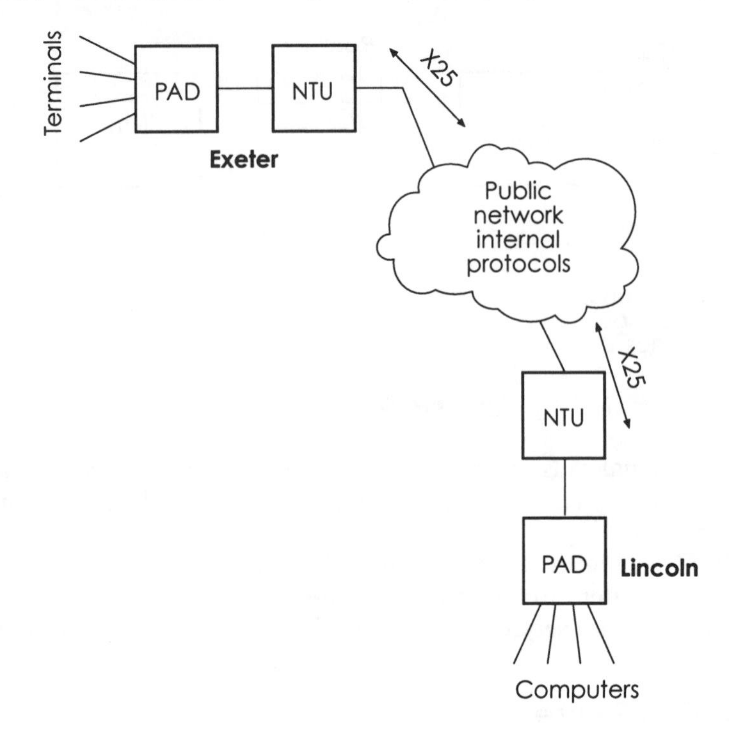

Figure 8.8 Public X.25

occasional data generators where the data quantity is also small the analogue system is attractive. In the context of the case study an appropriate typical PSTN connection might be made to shops and temporary exhibits where people are invited to join the organisation. For the organisation to offer immediate cover for new members it would be necessary to access the headquarters database. All that would be needed at the shop is a terminal and suitable modem as described in Chapter 6. Since only alphanumeric data would normally be entered at this stage the modem standard need not be complex. V22 would probably suffice. The analogue line from the modem will be connected through the nearest local telephone exchange and via the telephone exchange hierarchy to the nearest company network switching centre. The arrangement is shown in Figure 8.9. The operation of the system need not be entirely manual since both autodial and autoanswer modems are available. At the switching centre the PAD line is connected to the modem. An automatically generated textual sequence at the shop terminal would allow a 'one button' call set up through the modems and through the PAD to the new membership logon computer at head office.

The use of an analogue leased line is only justified where no connection can be made to the public digital network. The costs are comparable with digital leased lines and thus there is no real incentive to choose analogue connection deliberately. If an analogue leased line is the only option then a high speed modem standard such as V32 should be used, especially if the remote office were anything but a casual user of the kind described above.

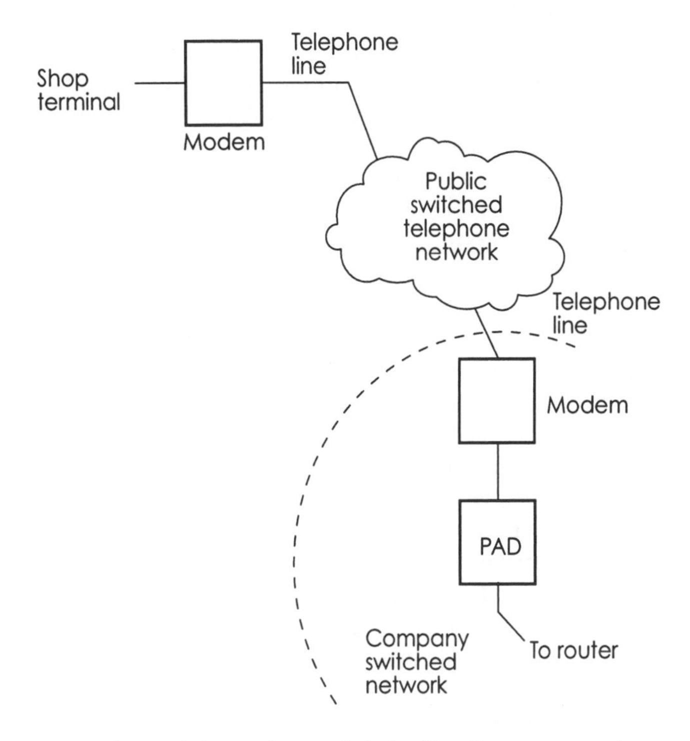

Figure 8.9 Analogue links to the Company net

8.4.5 Radio and satellite links

Terrestrial communication with mobile stations could be carried out using an entirely private network transmitting over licensed channels. Base stations and relays would be company owned and maintained. The base stations could be situated at the regional office sites and extra relays could be installed on rented sites either on high ground or on top of tall buildings. Although there is significant expense in this strategy, radio communication with mobiles is core to the company's business. The main advantage in radio network ownership is that the protocol is fully in the control of the company and also since the bandwidth is not shared there should never be a connection difficulty through congestion. A mixed data, speech and still image system would be supported by a private PAKNET operation. The details of PAKNET are given in Section 7.1.5.

There may well be areas where communications needs are relatively small, perhaps because of sparse population or because of geographical isolation. The company might consider that the expense of private radio net coverage in these areas is unjustified. It is possible that a public carrier could consider the area viable on the basis of carrying traffic associated with all forms of trade and private use. If this is so, then a cellular radio network might be available. The case study company could easily use telephonic or data communications of the V series modem type to communicate with its mobiles in this type of area. Connection to the cellular radio system is easily made through the public switched telephone network PSTN).

Geographical isolation is not the only problem facing an organisation seeking to achieve national coverage. There are areas which present geographical difficulties to radio networks using VHF (very high frequency – 30–300 MHz) and UHF (ultra high frequency – 300–3000 MHz) transmissions. These transmissions are largely 'line of sight' and real problems exist in mountainous or hilly terrain. Terrestrially, the problem can be solved only by passive or active relays strategically placed and in sufficient numbers. The population or traffic density expected for the case study business is unlikely to justify the expense and so a different remedy might be sought. Satellite communication (satcom) is recommended for quality links in these circumstances. Recent advances in satcom technology have produced very small and economic mobile ground stations. One satcom authority has developed a low powered terminal with an omnidirectional aerial. The prototype size of this station was 30 × 22 × 11 cm and the aerial is 8 × 5cm with the whole station weighing about 6 kg. The system allows for data communication at 600 bits/second, which is sufficient for telex style activities but not speech. Speech channels are available by satellite and could be provided if necessary.

Satellites are also useful for providing constant feedback about the positions of mobile stations. The NavStar satellite navigation system gives global positioning using cheap portable receivers which generate a standard format data signal carrying latitude and longitude information which could be returned via a terrestrial radio link.

Questions

1 An industrial organisation has design, marketing and production facilities distributed in the UK. The sites are : Greenock, Cardiff, Basingstoke and Norwich. Each site regularly generates large data files for transfer. Suggest a network specification using primarily X stream services provided by an external carrier. Justify your choices and indicate the need or otherwise for any switching centres, multiplexers or other devices.

2 Review the options and make recommendations to a project manager who has to create a local area network for general machine control in a commercial vehicle factory. Comment on the suitability of your recommendations for a drawing office environment.

3 A building society needs to communicate between its branches throughout the country. Assume that you have to explain the implementation options and write a short briefing note covering issues of traffic density offered and rehearses arguments concerning the choices that would have been made:

(i) analogue networks;
(ii) X21 leased lines at 64 kbit/second
(iii) X25 Packet switching.

4 Some network designers specify the use of a data PACX. What is this device and how might it be used in the creation of a private network?

5 What advantages might accrue from the use of ISDN links within a company private system?

9

Communications data processing

When public networks are used there is clearly a financial cost benefit in ensuring that no excess or redundant data is transmitted. Although the benefits are not financial, there are also advantages in a similar strategy for local area networks or links over leased lines. In these cases the saving is in bandwidth.

A data highway will have only a given capacity and transmitting redundant or unnecessary data is wasteful. When extra nodes are added to the network, saturation will occur earlier and expensive total upgrading, which could have otherwise been avoided, will be prompted. Implementing a useful processing function for ensuring that redundant information is removed from network messages requires an appreciation of the theory of information. Data communications processing certainly covers data compression but it also covers an increasingly important subject, the security of data. Data encryption is the technique used to make messages unreadable without extraordinary knowledge and effort.

9.1 Basic information theory

It is necessary to begin with the definition of the term 'entropy'. The definition is simple and is linked to the more well known use of the term in thermodynamics. In that subject entropy is a measure of disorder. Information theory uses the term to define the amount of information in a message. If a message is entirely predictable then it carries no information. If it is wholly unpredictable it is all information. Forms of order are predictable. Disorder is less so.

What is meant by predictable or unpredictable? Normally, a predictable message is one for which there is a high probability of being able to guess it in advance. Clearly, there is a link between entropy and probability. The entropy in a message is expressed as a negative logarithm of probability.

$$\text{Entropy} = -\log_2 \text{probability}$$

This is best understood by its graphical interpretation, as shown in Figure 9.1. If the probability of the next entity from the source is 1, i.e. certain or fully predictable, then the entropy is zero. If a binary source is assumed with equal probability of 0 or 1 as the next entity, then the entropy is 1. In this case 1 bit. Because of the uncertainty the entity cannot be signalled in less than 1 bit.

Imagine a textual source where a certain symbol or character has a statistical probability of appearing every 32 symbols. With a probability of 1/32, Figure 9.1 demonstrates that the entropy will be 5, i.e. 5 bits are needed to represent the inherent information content. Many text files are stored or transmitted over networks as ASCII characters of constant size 8 bits, irrespective of their probability of occurrence. The symbol in question will have an information content of 5 bits and a length of 8 bits if sent in this way. The 3 bit difference is redundancy. If a statistical analysis of the symbolic content of messages is carried out prior to sending them there will be an opportunity to reduce the messages to their pure information content.

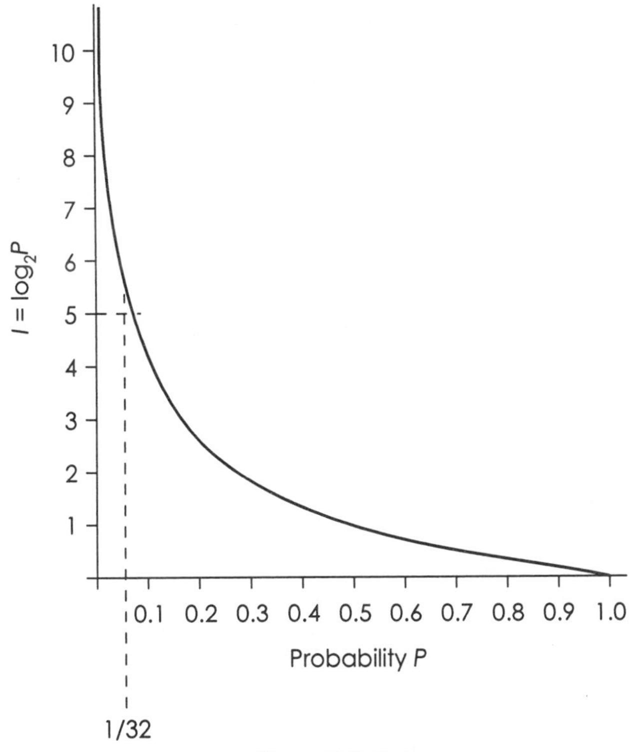

Figure 9.1 Entropy

9.2 Data compression

There are two main types of data compression:

- redundancy reduction
- entropy reduction.

Redundancy reduction incorporates methods that enable messages to have their information content identified. Once the information content is identified, the process is followed by a coding stage which seeks to allow transmission without redundancy. In this case no information is lost when the message is decoded and the overall strategy is lossless. Reconstruction is possible without error and this is essential where it is crucial to preserve the exact form of data, for example, when storing or sending computer programs or numerical data. This perfect reconstruction feature exacts a price in terms of the data compression that is possible. Although compression depends on the inherent entropy of the source, typically between 2:1 and 5:1 can be achieved. Although this sounds modest it is well worthwhile, provided that the execution time of the compression and decompression algorithms is not excessive.

Entropy reduction methods actually achieve compression by losing some information. This is acceptable only when no important data is lost. For example, images often contain far more information than the eye can interpret or than the display can portray. Obviously it is necessary to be able quantify 'acceptable loss' in order to use entropy reducing techniques.

In networking the most commonly used techniques are lossless and a number of different methods are popular. Even so, the advent of ISDN, the integrated services digital network, presages a time in the near future when speech, images and textual data will all be networked.

9.2.1 Run length coding

Appropriate only when data has a repetitive nature, run length coding might be used at bit or word level. Consider a data stream in which one character is repeated many times in succession, for example:

0,0,0,0,0,0,0,0,0,0,0,2,3,7,1,5,0,0,0,0,0,0,0,0,0,0,0,0,0,0,0,0

The multiple 0s might be encoded by <esc>,number of occurrences.
This would yield:

<esc>,11,2,3,7,1,5,<esc>,16.

Of course, only run lengths of up to 255 could be accommodated by a single 8 bit run length codeword. The compression achieved in the example is 32:9, just over 3:1. The <esc> character could not be permitted to exist as genuine data if this scheme is used. The problem here is that only the redundant runlengths are encoded. Consider the bit oriented case. If a sequence of bits were:

0000000111111100000000000001111000000000111111111100000000000000

then both the 0s and the 1s could be coded yielding:

7,7,13,4,9,10,14.

If the maximum run length were 16 then only 4 bits could represent the counts as $n + 1$; e.g. 0000 = count 1, 1111 = 16 and the sequence would be:

0110011011000011100010011101

A compression ratio of 64:28 or just under 2.3:1 would result. Clearly, the longer the run lengths in given data the greater the potential. Also, there is an implicit assumption that the first sub sequence is of 0s. The most appropriate data for run length encoding is graphics screens carrying line diagrams. The technique is often used as a pre-process for more sophisticated methods.

9.3 Huffman coding

The underlying concept involves variable length coding. A message is analysed in order to determine the occurrence frequencies of its individual elements. These elements could be ASCII characters. Those elements with the highest occurrence rates are allocated the shortest codes. This is best illustrated by example. Consider a message which contains symbols or characters with the frequency indicated:

a	s	d	f	g	h	i
22	11	7	33	28	67	61

The list is re-arranged in descending frequency of occurrence giving:

h i f g a s d

The two least frequent elements are then combined into a sub-group and their occurrence combined:

h	i	f	g	a	s	d
67	61	33	28	22	11	7
					18	

The next step is to combine the next two lowest occurrence elements or sub-groups i.e. 22 and 18. The process is continued and an inverted 'tree' is built up.

Each branch of the Huffman tree is allocated a '0' or '1'. Once the tree is complete the code for a particular character or element can be read off by concatenating the '0's and '1's as a branch is traced from tip to root. The scheme has ensured that the most frequent elements are on the shortest branches. In this case 'h' is encoded 00,'g'is 101,'s' is 1110. Consider the effect on a part of the overall message shown below:

h i h h g d i a s i i

If this had been sent as 8 bit ASCII characters the total number of bits would have been 88. In this Huffman code the binary message is:

00 01 00 00 101 1111 01 110 1110 01 01

a total of 28 bits. These codes could be sent as a continuous string of bits because there is an easy way to decide how to split up the string to decode the message. Starting at the beginning, the first bit is examined. If it is a '0' the first element is represented by a two bit code. If the first bit is '1' the very next bit is examined. If it is a '0', the code is three bits, but if it is a '1' then the next bit is looked at. If that bit is a '0' the code is three bits too but otherwise the code is four bits. No delimiters are necessary. To use Huffman coding as described it would be necessary to analyse the full message before it was transmitted over a network or stored. This may be impractical for some data sources and an alternative strategy is possible: adaptive Huffman coding.

9.3.1 Adaptive Huffman coding

In ordinary Huffman coding the coding tree has to be built on the basis of the occurrence frequencies of the message entities. This implies that the overall message is subjected to a statistical analysis prior to coding. Once the tree has been decided upon it will be used by the transmitting end but it must also be used by the receiver to decompress the message. Since the receiver is ignorant of the coding tree, details must be sent from the transmitter to the receiver ahead of the compressed data. This represents information additional to the message but the achievable compression of the message proper much more than compensates.

The idea behind adaptive coding is that the tree is not built first. It is built up as the data is transmitted. What is more, both transmitting and receiving ends construct the tree almost simultaneously, adapting which messages entities appear where in the tree as the process goes along. Provided that there is no data corruption on the link between the two ends then the tree evolves identically. There is no need to send tree details ahead of the message and so a better compression ratio results. There is a processing overhead because of all the shuffling of tree structure but this just means slightly slower compression and decompression rather than a higher transmission cost.

Although Huffman coding is an efficient technique it operates only on minimum entities like alphanumeric characters. It is unable to take advantage of the higher order associations between characters which sometimes occur in text.

9.4 Basic Ziv-Lempel coding

The working of this technique requires a 'flag' indicator to differentiate compressed data from uncompressed data. If the data were ASCII characters from the 7 bit set then there is a spare bit in each byte. This bit could act as the compressed/uncompressed flag. The original Ziv-Lempel system takes advantage of the repetition of words, phrases or parts of words. When a repeat sequence occurs a pointer is generated, pointing back to the previous occurrence. Added to this is a number indicating the length of the sequence. Consider a fragment of text such as:

> 'The brain drain has certainly aided Spain and the brain drain is certain to pertain to the UK.'

Of course, the example is contrived to illustrate the point. In longer texts repetition would occur more naturally but the uncompressed example has 93 characters. In compressed form the message looks as follows:

> 'The brain drain has cert/12/3/ly /6/2/ded Sp/26/4/and /46/16/is/44/7/ to p/11/7/11/3/41/4/UK.'

/nn/n/ represents two bytes carrying binary numbers. *nn* is the pointer back *nn* characters, including spaces. *n* is the number of characters to repeat. To understand this, imagine receiving the compressed message and decompressing each coded byte pair instantaneously. At the beginning of the message there are no compressed characters because none have been seen before.

'the brain drain has cert...' At this point the compressor recognises the character pair 'ain' and instead of sending the pair literally, it substitutes / back 12/3 characters/. It has made a small saving here, two characters sent for three in the text. The decompressor in the receiver counts back and repeats giving a text:

'the brain drain has certain... ' and carries on with the uncoded 'ly'. Although this is not a startling compression, consider what happens when the decompressor reaches /46/16/:
'the brain drain has certainly aided Spain and... '
Back 46 characters and the next 16 characters reads 'the brain drain', giving an updated recovered text:

'the brain drain has certainly aided Spain and the brain drain... '

This time the compression is good. A pointer and a length substitute for 16 bytes. Overall the original 93 bytes have been compressed to 60. Obviously, using two binary 8 bit numbers for pointer and length is limiting. The memory of the compressor can only be backward 127 characters. This is because in the scheme described the most significant bit of the pointer is used to indicate or flag that the byte is a coded byte and not a literal ASCII character. Similarly the length byte can cope with only 127 in this case. The length byte or word in basic Ziv-Lempel coding is said to define the active 'window' of memory on the text. Experiments with natural text and different sized windows, pointers and length words have shown that the optimal size for the window or 'sliding history buffer' is about 2 kbytes,which scales the pointer to 11 bits. Typical performance for an optimised system compresses natural text to about one third on average. Basic Ziv-Lempel coding is often referred to as the LZ-77 algorithm.

9.5 Ziv-Lempel trie coding

The deficiency of LZ-77 is that previous text is unusable once it beyond memory because it is further back than the pointer can reach. The remedy is to involve a much larger memory but using a special data structure called a 'trie'. A trie is illustrated by Figure 9.2 on the next page. As words are encountered, they are attached to a logical branch structure as shown.

Figure 9.2 is a diagram for the phrase 'famous five find favourite flavours'. Each word is specified by a pointer to the end of each branch. If a trie system has 32 000 k end nodes and they are pointed to by a 16 bit pointer (as in LZ-77, one bit is used to indicate coding), then the example phrase of 35 bytes will require 5 16-bit pointers and 4 literal space characters, a total of 14 bytes.

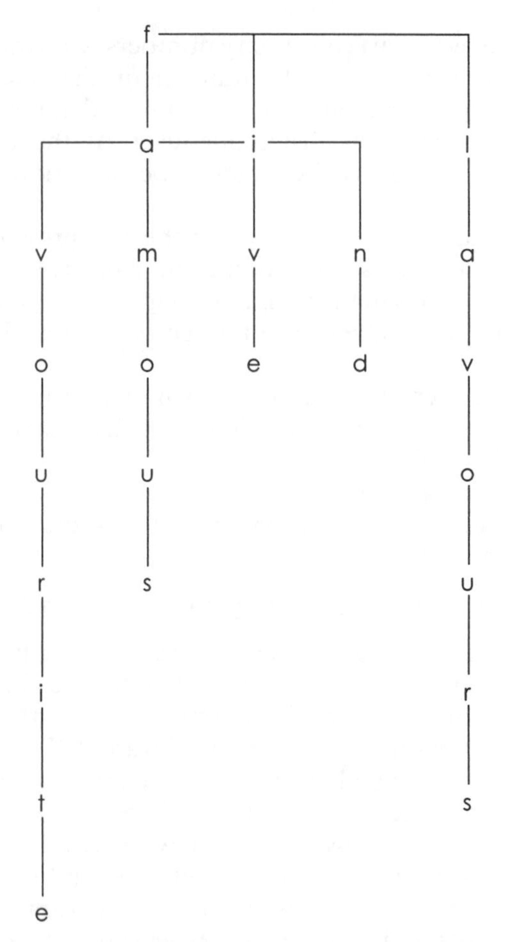

Figure 9.2 Ziv Lempel Trie

This method builds up a dictionary based on tries. It does this by adding new tries and new branches as new phrases occur. The method is known as LZ-78 and it does have a number of disadvantages to complement its otherwise good compression. The first problem is that sooner or later the number of possible node ends is exhausted. Once no more phrases or words can be added it becomes necessary to monitor the compression ratio being achieved. If the data stream changes its character then the trie based dictionary already being used may become inappropriate. If this happens the compression ratio will suffer and the best thing to do is to throw away the dictionary and start a new one. The second problem is that the receiving decompressor must have the exact trie dictionary model that is being used by the transmitter. It must also be managed, i.e. updated or thrown away, in exactly the same way.

9.6 Arithmetic coding

The idea behind arithmetic coding is that a string of entities could be represented by a unique number. This number is a fraction between 0 and 1 and the first stage in identifying the number is to analyse the string for occurrence frequencies of the entities making it up. In this respect the initial principle is similar to Huffman Coding. An example will illustrate this. Assume the need to code a simple word such as 'compresses'. Table 9.1 shows the entities and their frequency.

Table 9.1

Character	Probability
c	0.1
e	0.2
m	0.1
o	0.1
p	0.1
r	0.1
s	0.3

The sum of probabilities is 1 and each individual probability is allocated a place within the range 0 to 1. The exact way in which the allocation is made is not important but it must be consistent since the receiving decoder must use this same information. Table 9.1 lists the character entities alphabetically and allocation along the probability line 0 to 1 will also be made on this principle. Table 9.2 thus indicates a range for each character.

Table 9.2

Character	Range
c	0.0 to 0.1
e	0.1 to 0.3
m	0.3 to 0.4
o	0.4 to 0.5
p	0.5 to 0.6
r	0.6 to 0.7
s	0.7 to 1.0

The general idea behind coding is that a continuous 'narrowing' of the interval between the range progresses as each character of the word is coded. If the first letter of the word had been 'r', then the final codeword would be a fraction between 0.6 and 0.7. With the word 'compresses' the first letter is 'c' and thus it is known straightaway that the final codeword will be in the interval 0 to 0.1. Exactly where the codeword is in that interval is dependent on the letters following 'c'. Table 9.3, which is on the next page, shows how the final codeword is developed. The calculation is made as follows:

Interval(low) + Range(low) × Interval width = New Interval(low)

Interval(high) + Range(low) × Interval width = New Interval(high)

Table 9.3

Letter	Range	Interval width	New interval
c	0–0.1	1.0	0–0.1
o	0.4–0.5	0.1	0.04–0.05
m	0.3–0.4	0.01	0.043–0.044
p	0.5–0.6	0.001	0.0435–0.0436
r	0.6–0.7	0.0001	0.04356–0.04357
e	0.1–0.3	0.00001	0.043561–0.043569
s	0.7–1.0	0.000008	0.04356566–0.043569
s	0.7–1.0	0.000003	0.0435681–0.043569
e	0.1–0.3	0.0000009	0.04356819–0.043568217
s	0.7–1.0	0.0000000287	0.043568208–0.043568217

The word 'compresses' has been coded and is represented by any fraction between the final interval, i.e. 0.043568208 and 0.043568217. This usefully allows for 'approximate arithmetic', but normally the word is theoretically represented by the lower figure, 0.043568208. In binary notation such a fraction may need 32 or more bits, but if 32 were adequate for 'compresses' then a compression ratio of 10:4 or 2.5:1 would result. In common with Huffman coding, the method needs to analyse the input data stream and communicate a coding scheme to the decompressor. Also in common with Huffman coding, no advantage can be taken of relationships between characters. For the decompressor to identify the end, an end of message code must be transmitted. Although it would seem for long messages that a very long fraction might cause an arithmetic overflow, it is not the case. The reason for this is that, once a particular digit is set and the final fraction is not yet reached, that bit stays set and may be transmitted or stored. A subsequent shift of the fraction register may take place, allowing more bits to be handled.

9.7 Entropy reduction

Networks will need to carry image data when accessing databases which store not only text but also illustrations of many kinds. This already occurs in networks used by publishing houses to compile texts containing photographic material. Internetworking will become a necessity for exchanging data of this kind and the ISDN already offers a service. The real problems in networking photographic images stems from the very large data content. A single monochrome image with 256 grey scales would require a byte for every picture element or pixel. A typical photographic image might comprise of over a million pixels and even using the full bandwidth of Ethernet only about 6 to 8 full pictures could be transmitted each second. The problem is much worse when considering an ISDN B channel offering only 64 kbit/second. Using the full bandwidth, each image would take about 15 seconds to transmit. In a real world network, the full bandwidth is unlikely to be available to a single user and so it can be appreciated that image data compression is an important issue.

9.7.1 Transform coding

Transform coding is a useful technique for dealing with compressing grey scale image data. A characteristic of raster scanned *n*-valued image pixels is that they are usually strongly correlated. The purpose of transform coding is to use a linear transformation in which the samples are mapped into a transform space. This process results in a set of samples that are more independent. The transformation does not of itself provide data compression but the processes that do, such as quantisation, can be more effective with more independent samples. The method is classified as entropy reducing because the term 'entropy' means the average information content of a message and processes like quantising will always eliminate some data as being negligible. After quantising, some data has been lost, that is some information has been lost. Another way of saying this is to state that 'entropy' has been reduced. The trick in using these techniques is to know how much data can be lost without significantly damaging the message. Such techniques are suitable for images, which often carry far more information than is ever usable.

To understand transform coding it is necessary to consider a function of time such as a line of a raster scanned image. Taking any sample as a starting point, the next sample will not be entirely independent. That is, a sample in the vicinity of the previous one will most probably be close in value rather than wildly different . This indicates that there is correlation between them. The purpose of a 'transform' is to decorrelate the samples.

9.7.2 The Fourier transform

In general the Fourier transform allows a signal that is a function of time to be represented in the frequency domain. To appreciate that the move into the frequency domain achieves a decorrelation it is necessary to think about the nature of the samples in each domain. A time domain signal is likely to be a series of sample values, perhaps of voltage. As previously explained a given sample value will to some extent be dependent on the previous one. The Fourier transform operates on such a signal to produce a 'spectrum ' which shows the frequency components necessary to synthesise the original function of time.

The frequency components in the spectrum are more independent than the original time domain samples. This may be confirmed by imagining the result of removing some samples. Removing some time domain samples gives rise to 'gaps' in the time domain signal. Removing some frequency components does not give rise to gaps when the time domain signal is recovered by the inverse Fourier transform.

Time domain signal samples may be modelled as a series of weighted Dirac functions :

$$f(t) = x_0\partial(t) + x_1\partial(t-T) + x_2\partial(t-2T)$$

The Fourier transform of this function is :

$$G(jw) = x_0 + x_1.e^{-j\omega t} + x_2.e^{-j2\omega t}...$$

This transform of a sampled data signal is generally known as the discrete Fourier transform (DFT).

Specifically, the DFT is given by the expression

$$x(K) = \sum_{n=0}^{N-1} x_n.W_N^{KN} \qquad K = 0,1,\ldots,n-1$$

where $X(n)$ is a signal sample value and $X(K)$ are frequency domain values

$$W = e^{-j2\pi/N}$$

Direct computation requires N^2 complex additions, which can produce a large overhead which has prompted the development of the fast Fourier transform (FFT).

The basic idea behind the FFT is that the summation over N samples can equally well be achieved by a combination of summations over $N/2$ samples. In the case where N is a power of two, the sample grouping principle can be applied repeatedly, culminating in a final summation of only two values.

There is an algorithm that accomplishes the basic operation needed. This algorithm is called the Radix-2 DIT (decimation in time). The technique breaks down the DFT into two smaller transforms, thence into two smaller transforms and so on. Eventually, the DFT is broken down into a number of

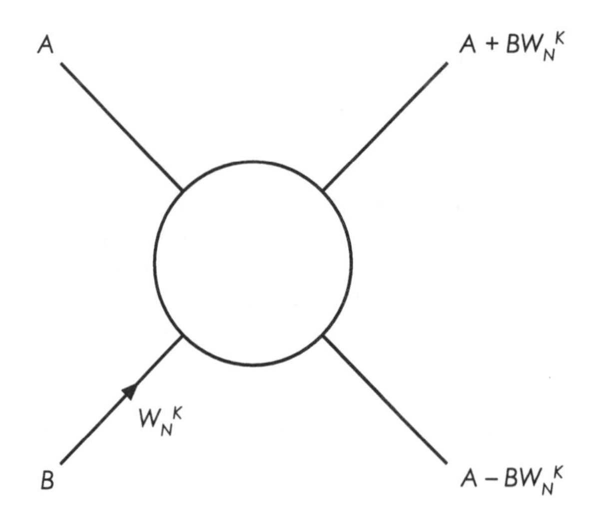

$A = a_1 + jb_1$

$B = a_2 + jb_2$

$W_N^K = \cos(2\pi_K^N) - j\sin(2\pi_K^N)$

Figure 9.3 FFT butterfly

two point transforms. A two point transform is known as a **butterfly**. The origin of the name derived from the configuration of the signal flow diagram that may be used to describe it is shown in Figure 9.3.

The operation accepts two complex words, A and B and delivers two transformed words. Any system attempting FFT action will involve a complex multiplication and two complex additions. The complex multiplication may be achieved by four real multiplies and two real additions. Complex addition can be provided by two real additions.

This leads to a total number of computations required to implement the butterfly of four multiplications and six additions. The total number of butterflies required to execute an N point transform is :

$$N/2(\log_2 N)$$

Thus a 1024 point operation is achieved by 5120 butterflies, and the computational advantage increases with N.

9.7.3 Transfer coding and quantisation

Merely transforming into the frequency domain does not generate a compression of itself. The transformation produces strong compression potential with all the high valued coefficients bunched together, followed by a long run of similarly valued ones. This long run consists of zero or near zero valued coefficients which may be allocated zero.

These contribute very little and may be ignored or compressed because of their redundancy. Zero valuing small but non-zero coefficients will create some small distortion when the data is reconstituted. The amount of distortion acceptable for a given system will determine which coefficients must be considered significant, and which may be zeroed. The allocation of bits to coefficients either zonally (only allocating significant values to a particular portion of the frequency spectrum – in this case equivalently low pass filtering) or by threshold (zeroing all values below a given amplitude) is a function of quantising.

9.8 Case study consideration for compression

The first consideration is whether data compression should be applied or not. The specification for the network requires client-server relationships involving programs and data. In the examples the data was assumed to be text but executable program files can exhibit similar properties. Perhaps the easiest example of an equivalent repetitive phrase is the existence of program macros. Clearly, executable files are less amenable to compression because instruction words or operators are smaller entities than text words and they are modified by the operand. How many times can it be expected that the same instructions operating on the same memory addresses or ports will be encountered? Certainly it will be less than the occurence of 'the' or 'and' in a text data file. Different compression algorithms are more or less effective in a mixed text and program environment. The results of extensive testing by several authorities are shown in Table 9.4

Table 9.4

Algorithm	Overall average compression per cent
Huffman coding	32
Adaptive Huffman coding	35
Ziv-Lempel LZ-77 coding	51
Ziv-Lempel LZ-78 coding	48

The LZ options in Table 9.4 are best performances based on optimising the index/length word sizes. They are also variants of the basic algorithm which seek to minimise processing. What is not apparent from the data is that there are speed differences which might affect selection. LZ-78 implementations tend to be slower than LZ-77. Huffman coding options are faster than either. Since the specification calls for office type activities on the local area network, speed is not as critical as it might be with, for instance, real time control. Links via the public data network or satellites are more likely to be limited by the available bandwidth than compression latency.

9.8.1 Throughput and latency

Latency may loosely be thought of as delay. If a compression algorithm is slow and a network is operating a stop and wait protocol as described in Chapter 4 then, before an acknowledgement is sent and the next data frame dealt with, there could be significant delay. The throughput of the system would suffer. Windowed protocols allow overlapping of the compression/decompression delay with the sending of further data frames. Since previous design decisions have selected a windowed protocol, throughput should not be seriously affected by the choice of a moderate speed algorithm.

9.8.2 Effects of transmission error

When compression systems use adaption, trees or tries they are extremely sensitive to error. If an update is corrupted then the evolution of transmitting and receiving structures diverge with disastrous results. Compression based on static statistics is less vulnerable but produces poorer overall results. The decision depends upon whether an error free link is practically guaranteed at the data frame level. The previous choice of LAPB as a frame protocol is sufficient such guarantee.

9.9 The selection

9.9.1 Textual data

The combination of layer 2 decisions result in no serious effects due to compression algorithm latency or to transmission errors. This being so the decision to use LZ-77 or an optimised variant would appear to be entirely reasonable.

9.9.2 Image data

The CCITT standard H264 for videophone images uses transform coding' to achieve powerful compression. CCITT and ISO recognised the importance of a compression standard for images and set up a Joint Photographic Experts Group (JPEG) to define both a baseline entropy reducing method and a lossless standard. The JPEG entropy reducing standard is the chosen technique. The method has three stages: transformation; quantisation; and lossless compression.

9.9.2.1 JPEG transformation

Although the Fourier transform works reasonably well it is possible to obtain better performance by using a related transform, the discrete cosine transform (DCT). The DCT may be obtained from a signal by first applying the fast Fourier transform (FFT) and then multiplying each of the frequency domain points obtained by an exponent. The full procedure is defined by Figure 9.4.

The FFT was explained as a single dimensional transform in Section 9.7.2 but two dimensional transforms are easily implemented. In the JPEG model, the two dimensional approach is used and the image is treated as an $N \times N$ matrix.

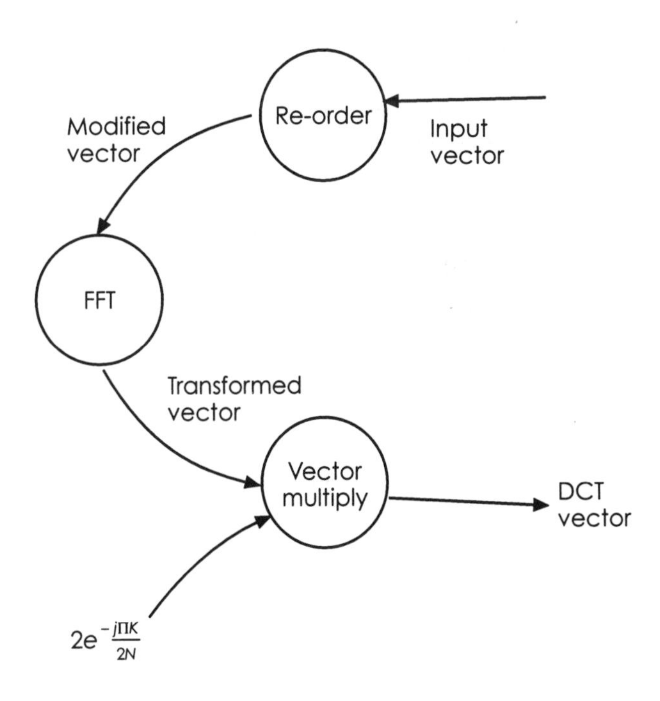

Figure 9.4 The Discrete Cosine Transform

9.9.2.2 JPEG quantising

The transformation process produces a matrix result. The JPEG quantisation strategy is to multiply the result matrix by a special quantising matrix. Each element of the result array is divided by the number in the peer position in the quantising array.

Quantised value = DCT(i,j)/quantarray(i,j)

The result of the division is then rounded to the nearest integer value. The purpose of quantisation is to reduce the number of bits needed to fully represent an entity. The values in the quantising array may be chosen to achieve different ends. One array might result in a minimising of the error between pre- and post- encoded images. Another might seek to produce an effect that was most pleasing to human perception. Surprisingly, the two examples given do not produce similar quantising arrays. The use of arrays offers the opportunity to select quantising arrays appropriate to different qualities in the decoded image. To see how this system works consider the small DCT result array and a corresponding quantising array as shown in Figure 9.5

After quantising, a large number of the elements will be zero or very low valued. Generally, only those in the top left of the output array will have significant values. The overall process seems to have concentrated all the non-zero values.

8	16	19	22	26	27	29	34
16	16	22	23	27	29	34	37
19	22	26	27	29	34	34	38
22	22	26	27	29	34	37	40
22	26	27	29	32	35	40	48
26	27	29	32	35	40	48	58
26	27	29	34	38	46	56	69
27	29	35	38	46	56	69	83

Figure 9.5 Quantising array

9.9.2.3 JPEG coding

To transmit the data in an efficient way it becomes necessary to read the array elements in a specific way. Figure 9.6 illustrates the zig zag reading pattern. As a result of this method the non-zero values come first, followed by all the zeros. This makes an ideal sequence for run length coding. Implementers are

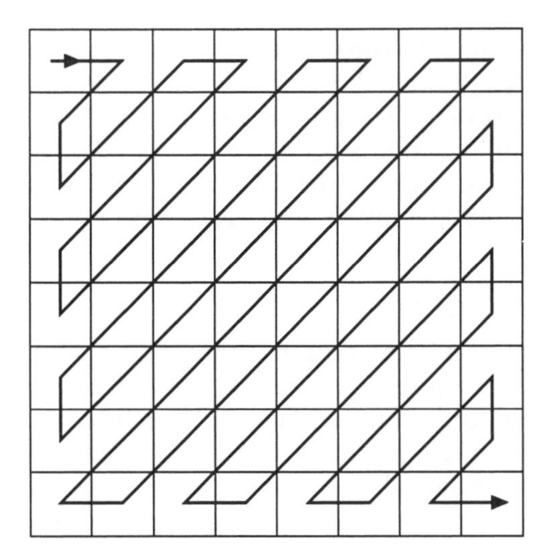

Figure 9.6 Zig zag coding

also allowed to use arithmetic coding or Huffman coding in order to create a compression.

9.9.2.4 JPEG decompression

To recover a coded image the final transfer coding method must be reversed. Once this is done the resulting array must be written back in a reverse zig zag. Then, the same quantising array used in compression is used. This time each element multiplies its corresponding element. Finally, an inverse DCT will produce the original image pixel array, albeit subject to minor and hopefully acceptable distortion.

9.10 Data encipherment

Computer networks are vulnerable to electronic eavesdropping. Commercial devices normally used for monitoring the traffic on the network and verifying the correct action of the protocol can just as easily allow unauthorised persons to read the data in the data frames or packets. Local area networks extending over factory sites or university campuses have many little corners where a tapping could be made. In the case study unscrupulous people could become aware of the lonely location of a broken down car with a single occupant. One way of making it very difficult to obtain such data is to encipher the messages. Encipherment is alternatively known as encryption and it can be carried out as a layer 6 function. Cryptography has its own jargon born of a long history. The unencrypted message is called the 'plaintext'. The output of the process is called a 'cryptogram'. Strictly speaking cryptography is the business of devising or creating ciphers. Cryptanalysis is the business of breaking them.

The long history of ciphers and codes has meant that many options have been developed. The earliest were trivially easy to break and even some of the later attempts can be quickly broken if a full statistical knowledge of the frequencies of individual letters, digraphs (two letter groups) and trigraphs (three letter groups) is known. In addition some contextual information may well be possessed by the cryptanalyst and this can help. A simple example might be the logon sequence at the beginning of the access to a host computer, e.g. words like 'password' and 'username'.

9.10.1 Substitution Ciphers

With substitution ciphers a letter, digraph, trigraph or word is substituted by another letter, digraph, trigraph or word. At the letter level a new group of letters might be aligned under the message and the individual letter substitutions made. Rather than use a single substitution group a whole range of possible groups might be available. Say there were ten possible groups and a switch was made with each letter. The new groups used might be numbers 28159364. Knowing the whole range of groups and knowing the key 28159364 will enable a recipient to decrypt the message. Although this sounds sufficiently complex to foil the cryptanalyst it is not. Much more security is obtained if the key is actually longer than the message and is used once only before the key is changed. This 'one time key' is unbreakable if the key is chosen to be a truly random number.

There is a problem with unbreakable ciphers. Many of the world's governments, including the USA, are not happy with their use. This conflicts with the needs of commerce and industry and a compromise is often arrived at in which the permitted ciphers are almost unbreakable. This means that they are vulnerable to cryptanalysis but not to those possessing normal resources or computer power. An example of this will be discussed after the other main encryption method has been described.

9.10.2 Transposition ciphers

With transposition ciphers substitutions are made. The original message entities are merely jumbled according to a key. To understand this imagine a key word 'signal'. The message to be encoded is:'The password was a big secret'. First the key is written down and the alphabetical appearance number for each letter is noted. So, for 'signal', s is the 19th letter, i is the 9th letter, g is the 7th letter, n is the 14th, a is the 1st, l is the 12th. The message is written in 6 letter rows (the length of the key) underneath. The encrypted message is read off by reading the columns in a columnar transposition, lowest column number order as follows:

19	9	7	14	1	12
s	i	g	n	a	l
t	h	e	p	a	s
s	w	o	r	d	w
a	s	a	b	i	g
s	e	c	r	e	t

The encipherment yields:

adieeoachwseswgtprbrtsas

It is very important that the keyword does not have any repeated letters or unnecessary vulnerability is the result.

9.10.3 A Product cipher

Product ciphers combine the use of transpositions and substitutions, perhaps many of each, between the plaintext and the ciphertext. Such a scheme can be very demanding of any cryptanalyst, even with plenty of computer power available. This class of cipher will be explained with an example which is a heavily used system. It is called the 'data encryption standard' or DES. Messages are dealt with as blocks of 64 bits of plaintext. The key size is 56 bits, which has been reduced from an original design key of 128 bits. The reduction was made at the insistence of a US Government Agency. It is interesting to note that the original key size would have made the cipher very strong indeed.

9.10.4 The data encryption standard

It will be clear from previous discussion that the traditional approach to strong security involves long, one time keys used with a relatively simple algorithm for encryption. An alternative is to use a relatively short key but make up for this by using a complex cascaded algorithm. The second approach is not as secure but the sheer magnitude of the task of determining a convoluted algorithm will deter all but the most committed cryptanalyst. The DES standard is limited to a 56 bit key and so must compensate by the complex cascading of simple transpositions and substitutions. The overall set of processes may be implemented by software procedures. The functionality of transpositions and substitutions is illustrated by Figures 9.7(a) and (b). Transposition is simply jumbling of individual bits but, with substitution, a code converter is used which takes an input binary word, translates it into a 1 of *n* code, jumbles the position of the single '1' state and performs an inverse translation, selecting a different binary word in so doing. These transposi-

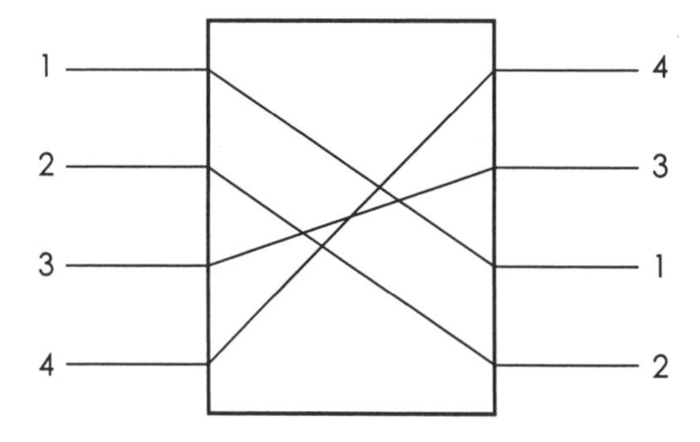

Figure 9.7 (a) Transpositions

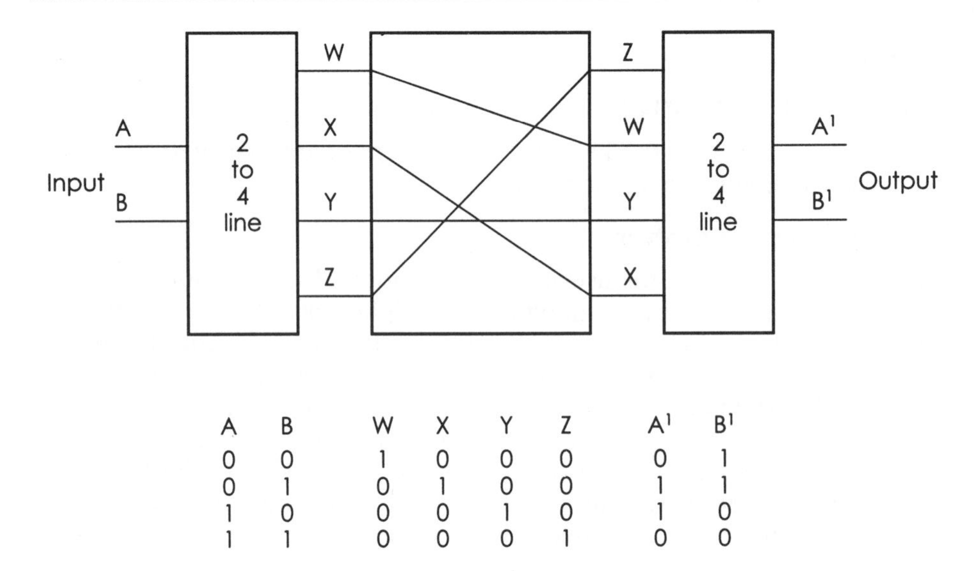

A	B	W	X	Y	Z	A'	B'
0	0	1	0	0	0	0	1
0	1	0	1	0	0	1	1
1	0	0	0	1	0	1	0
1	1	0	0	0	1	0	0

Figure 9.7 (b) Substitutions

tions and substitutions are performed serially in a cascade of operations as shown in Figure 9.8. The basic principles are used by the data encryption standard (DES).

The text to be encrypted is broken down into 64 bit portions and all the bits are transposed. The penultimate step requires that the first and last 32 bits are word-wise transposed. Thus bit 63 becomes bit 31, bit 0 becomes bit 32, bit 32 becomes bit 0 and bit 31 becomes bit 63. The last step is the exact inverse of the first transposition, but in between there are 16 further steps. The 16 internal steps are all identical except that they use different versions derived from the 56 bit key. Each internal step operates on two 32 bit halves simultaneously. Part of the difficulty is reconciling a 56 bit key with 32 bit entities. This is overcome in an ingenious way. One of the 32 bit halves is first transposed to produce a new 32 bit entity and then, according to a fixed rule, certain parts of it are duplicated to form a 48 bit entity. A 48 bit function of the 56 bit key is then exclusively ORed with the extended original 32 bit input half. The 48 bit exclusively ORed output is broken down into eight, 6 bit entities which are then dealt with individually. They are subject to a

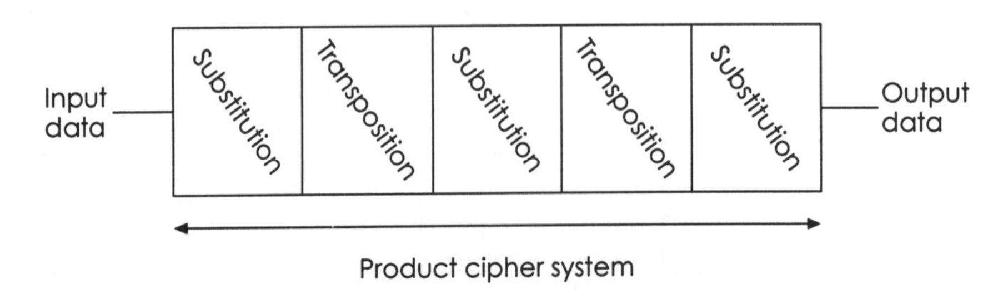

Figure 9.8 Cascaded processes

substitution in which the 6 bit inputs are mapped into 4 bit outputs. This step sounds complex but is, in reality, simple. A 6 bit input could assume any one of 64 possible states. A 4 bit output could assume any one of 16 states. To match input and output requirements, every output is generated by four input states. Different inputs can generate the same output. The full output of the system is eight, 4 bit words. In this way, a 32 bit input creates an enciphered 32 bit output and the whole process described is repeated 16 times in total. Each iteration uses a different key derived by transposing the 'official' key according to another fixed rule. It only remains to explain that the process has been applied to only 32 bits of the 64 bit original input. What about the other 32 bits? Each iteration simply 'passes through' the other 32 bits but takes care to swap 32 bit halves so that the whole 64 bits has been subject to encipherment after the first two iterations. Figure 9.9 attempts to capture the process as a flowchart. Decryption is relatively simple and involves the inverse of the total process.

Even though the algorithm is complex it is apparent that, given the key, a cryptanalyst could break the cipher. This is because the overall process is in the public domain and is by definition a standard. The question then arises as to how users deal with key distribution. Either all communicating parties must have agreed keys based on daily, weekly or monthly changes or a new key must be signalled to everyone prior to its adoption. Clearly, there is real danger in signalling the key. Eavesdroppers might easily obtain everything they need to know. Discounting ideas of distributing keys by courier, a time consuming and out of date notion, how might the job be done?

9.10.5 Key distribution: puzzles

The principle here is that a secure link between two correspondents is established by an initial 'in the clear' dialogue. During the dialogue, the initiator of the link tells the receiving end that there follows a large number of encrypted 'puzzles'. The definition of a puzzle is that it is an encrypted plaintext with a special composition. It is perfectly acceptable for the special composition to be defined in the initial dialogue. The puzzle plaintext might consist of a message beginning with a characteristic sequence which could be as simple as 64 consecutive 0s. At some point in the puzzle plaintext must be a unique puzzle number and a 56 bit key (assuming a DES key). Each key would be different and a large number of puzzles would be signalled from one correspondent to the other. Typically at least 20 000 puzzles would be signalled.

The fact that the recipient knows the composition of the puzzles would enable him to recognise when one had been correctly decrypted. The originator of the puzzles invites the recipient to choose a puzzle at random and break the encryption by trying every single possible key until the correct format puzzle plaintext is recognised. This method is generally known as breaking the code by brute force. For DES all 2^{56} key possibilities would have to be tried. Sometimes, the task is made easier by using a key with a lot of trailing zeroes. If the last 32 bits of the key were zero, the search would then involve only 2^{24} possible keys.

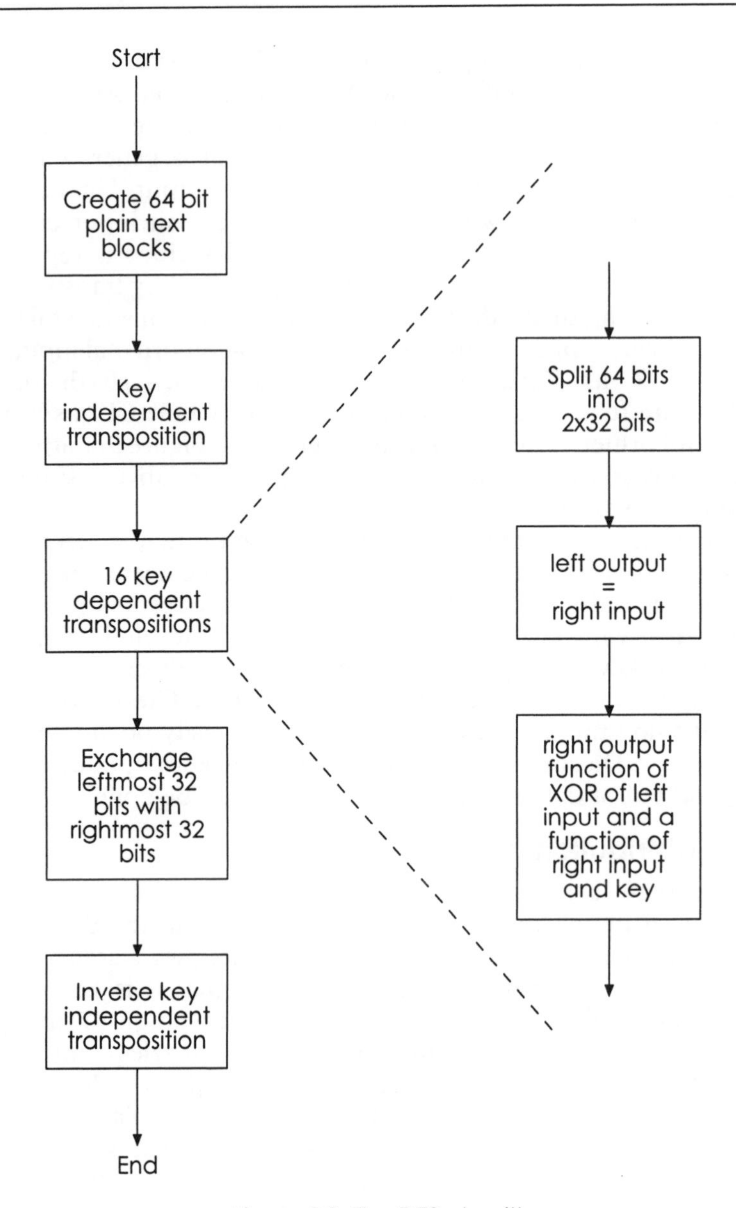

Figure 9.9 The DES algorithm

The point is that the intended recipient breaks the code by brute force and then tells the originator the number of the puzzle that has been cracked. Both then know the key that they will begin to use in sending encrypted data. Any eavesdropper will also know the composition of the puzzles. He will not know which of the puzzles has been chosen and broken by the recipient because the puzzle number is encrypted within the puzzle. Whereas the intended recipient only has to break one puzzle, possibly taking an hour or two, the eavesdropper has no choice but to attempt to break all of the puzzles until he discovers the one with the puzzle number in use. On average this means breaking half the total number of puzzles. Typically an eavesdropper will take 10 000 times as long (assuming 20 000 puzzles) as the intended recipient. If the intended recipient takes 2 hours to complete his task and establish a secure link then the eavesdropper working at the same speed as in this case would take, on average, over 2 years to find the correct key. Of course, this is a statistical figure and a fortunate eavesdropper might chance upon the code more quickly. The chances would be slim if the correspondents decided to repeat the process and have new keys daily.

9.10.6 Public key cryptography

The term 'public key' implies that the key is public knowledge. The intuitive notion that keys must be kept secret is not necessarily true. The basis of the thinking is that it is possible to define an encryption algorithm and a decryption algorithm which when used sequentially result in the original message being exactly reconstructed but which are such that it is extremely difficult to derive the one from the other. Although the details of such methods are outside the remit of this book, these methods are based on encryption and decryption algorithms that can be linked by a process of factorising very large numbers, typically in excess of 200 digits. This kind of problem is well known for being very difficult and it has been calculated that the job could take millions of years for an eavesdropper, even assuming that the highest computational power currently available is used.

9.11 Case study consideration for encryption

Criminals may be able to identify the geographical position of a lone broken down motorist. Eavesdropping is relatively easy with cellular radio links and for those determined enough it is also easy to tap a company's local area network. Often, rescue services can be delayed for a few hours and thus the use of data encryption is sensible. The need is not for a fully secure system since the information will have a lifetime of only a few hours at most. The DES standard would certainly be good enough for this task. Key distribution need not require special arrangements since rescue units return to a depot at least once a day. A new key could be entered daily from a codebook which is retained at the depot. A significant advantage of choosing the DES standard is that chip sets are readily available that implement stream encryption so that the DES encryption process is performed in fast real time mode.

Questions

1 Pattern matching and substitution may be used as part of a data compression scheme. Justify this statement and suggest a possible classifier. Further comment upon whether the suggested process is noiseless or entropy reducing.

2 The national maps authority wishes to reduce the network bandwidth used to transfer maps from servers to mapmakers' local terminals. Given that the authority is flexible on how maps may be stored, suggest a storage format that would facilitate data compression. Further develop a coding scheme to demonstrate how compression may be achieved.

3 What is the underlying purpose of using a transform in transform coding? Explain how compression is achieved in the overall transform coding process and define how the JPEG standard differs from the simplest scheme.

4 Differentiate between Ziv-Lempel algorithms LZ-77 and LZ-78. Identify the issues affecting the relative performance of these algorithms when used in compressing textual data.

5 Explain how the data encryption standard might be implemented in other than a block cipher operation.

10

Network management

10.1 Introduction

Network management is many faceted and applies at different stages in the life cycle. During installation, planning and facilitation are critical. Once a network is operational, the administration issues are:

- user access rights
- system security
- fault management
- fault recovery
- accounting services.

Effective monitoring of the network requires equipment and instrumentation to probe activity, quantify traffic patterns and execute diagnostic procedures. When networks become part of a corporate scene it becomes essential to ensure that the whereabouts and integrity of corporate data are guaranteed. Estimates derived from the top 100 US companies indicate network down-time costs approach $30 000 per hour, with average disabilities lasting 4.9 hours. When networks become larger than 30 users or so the network manager is unlikely to remember all the details of the cabling, hosts and users' machines. For them to continue to be effective, policies must be developed.

10.2 Facilitating installation

A sensible work plan involves an initial survey of proposed routing for the network data highway. Some of the features to be avoided are: hazardous environments; and cable bending radii in excess of that permitted by the cable manufacturer. Hazards may be electrical or physical. Highways should not be run in common trunking with supplies to inductive machines or adjacent to flourescent lamps. To aid technicians during the installation procedures, instruments are needed that are capable of measuring electrical continuity and impedence mismatches likely to give rise to electromagnetic wave reflections on the highway. It is good practice to provide a small laboratory network to enable each of the transceivers and repeaters to be tested independently of the full network prior to installation. If fibre optics are being used it is essential to have good quality jointing equipment. Appropriate cable or fibre pulling kits are essential to avoid damage during cable 'pull through'. The network layout should be designed and updated using a CAD system which can be annotated and modified to suit all the problem areas that are always encountered. If any repeaters or other devices are part of the network, their location should be clearly marked with permanent notices, especially if the devices is hidden in a liftshaft or behind panels. All cables should be marked during installation with numbered heat shrink sleeving. Other methods, especially those involving paper labels or taping should be avoided as they age badly, become dirty or unreadable or sometimes even lost.

10.3 Monitoring the network

Monitoring may be done at a number of levels. One type of monitoring uses an instrument which is literally 'tapped' into the data highway and non destructively intercepts all data frames. The instrument, known as a protocol analyser, enables the frame dialogue to be recorded and thus allows detailed analysis of traffic densities, sources and destinations, traffic profiles and statistics. The protocol analyser may provide a number of displays but probably the most useful in continuous monitoring of the net is the so called 'skyline' diagram, which is illustrated by Figure 10.1.

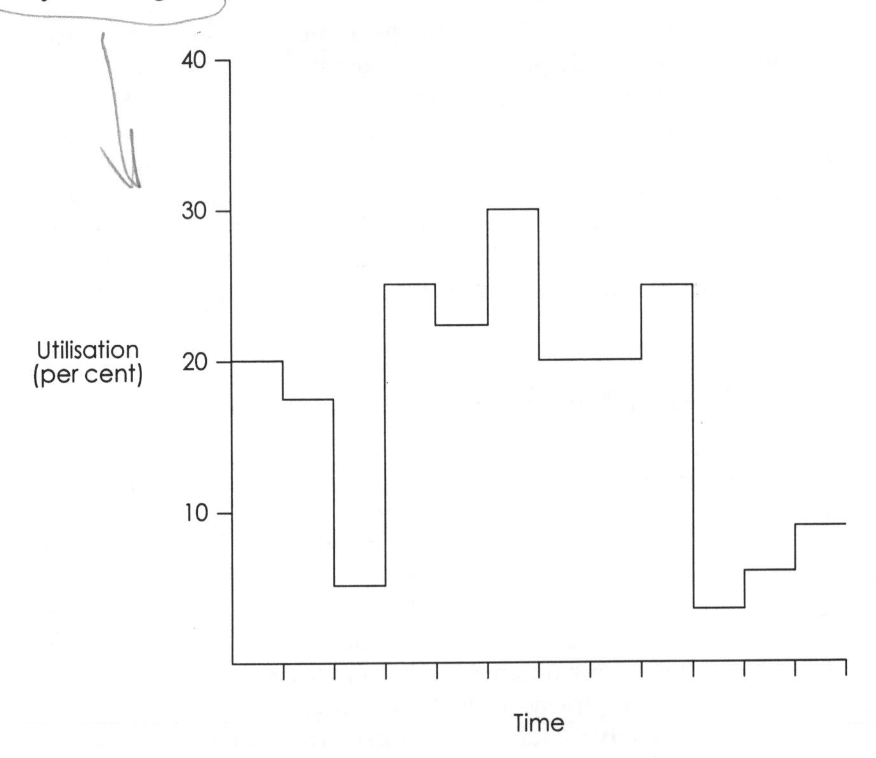

Figure 10.1 A skyline diagram

Analysis functions normally provided include the types of errors that are occurring. Typically, these errors might be bit errors, parity errors, framing errors, and burst errors detected by faulty CRC. The protocol analyser can be used in a 'baby sitting' mode. This is because it can operate as a data recorder with a 'first in – first out' equivalent shift register action together with a defined trigger capability. The trigger event, which could be the detection of an error, causes the recording to stop. The stop, usually after a short delay, allows the close scrutiny of events prior to, during and after the error. One very useful feature provided by some protocol analysers is node emulation. The analyser can be programmed to behave in the same way as a planned additional node. In this way the effect of the addition can be tested without investing in a full prototype node.

A number of low level errors may be detected and it is useful to understand their origins and to appreciate the significance when large numbers of errors emanate from a single cause. Generally, low level errors are classified as: runts; jabbers; misaligned frames and collisions.

10.3.1 Runts

Runts are data frames that have suffered destructive interference and are then shorter than the minimum size allowable for a data frame in the system. Runts are most frequently created by collisions in contention nets. Often a runt is so disrupted and shortened that it has no remaining source address data field. For this reason the origin of the frame before it became a runt or equally the node that caused the collision remain unknown. Network managers are generally more concerned with monitoring the number of runts. Large numbers of runts indicate problems.

10.3.2 Misaligned frames

Many communications protocols define the minimum size for a data frame. This minimum and other legal sizes are usually a multiple of a certain number of bits, typically eight. If a frame is received and the CRC is in error and furthermore the frame size is not divisible by eight, then the chances are that the frame is misaligned. A prime cause of misaligned frames has been found to be repeaters or similar devices, which have subtracted or added bits to the frame in passing it through.

10.3.3 Jabbers

Whereas runts are short frames, jabbers are longer than is legally defined for the network. The sources of jabbers are often traced to driver faults in the node interface or to poor connections. Some networks include nodes or repeaters which actively look for jabber sources and may attempt to isolate or cut them off from the rest of the network.

10.3.4 Collisions

In a contention system collisions are normal and problems have been minimised by access methods such as CSMA/CD. Some collisions may occur and it is as well to monitor them in order to be able to correlate their occurrence with other factors such as network utilisation. One network manager discovered that the highest number of collisions was strongly correlated with the lowest utilisation of the network. This was due to the fact that the lowest utilisation was when the most probabalistic accesses took place. Periods of high traffic on his network corresponded to orderly and predictable archiving.

10.3.5 Timing measurements

Networks are dynamic and tend to grow. The growth will be in numbers of nodes and resulting traffic increases. In addition growth usually results in greater geographical distribution or dispersion. To be sure that a network retains a satisfactory performance there are two parameters that should be monitored. These are channel acquisition time and network response time.

Channel acquisition time is a measurement which globally includes all of the detailed media access control elements such as collisions, deferrals and back offs. If the channel acquisition time degenerates it is time to analyse the reason. Channel acquisition time is usually measured by issuing a large number of messages and working out the average time taken to dispose of the data frame.

Network response time measures the time necessary to achieve a round trip communication. That is, to issue a message and receive a reply. This is particularly useful when local area networks are linked by a wide area network or long distance link. Response times for local nodes may be compared with response times of remote nodes at the other end of a link. In this way network managers can quantify the time costs of long distance links.

Another aspect of monitoring concerns not just the mechanics of network data transfers but the parameters necessary to manage the network effectively on a day to day basis. These are typified by:

- storage management
- systems management
- network management
- resource management.

To understand the need for these features it is necessary to imagine a network of significant size in a large organisation with a formalised IT service. The keyword is 'service' and the type of services required will not differ substantially from what was expected in the days of mainframe dominance. The only difference will be that computers are more ubiquitous and on everyone's desk.

10.4 Storage management

The main issues in storage management are disk backup and tape storage and control. The servers within networks usually carry considerable numbers of applications and for many organisations, large amounts of data. The data is a valuable resource to the company and cannot be allowed to be lost under any circumstances. The applications are usually supplied on a transferable medium but they take significant time to install, tune and configure. It is sensible to archive all material and to do it in an automated way. Typically, archiving should be triggerable by time or by the recognition of a trigger threshold in terms of a defined percentage of storage already used. In this way data is safely archived before any uncontrolled situations occur. Some operating systems still do not cleanly handle an event like running out of disk storage in the middle of storing a file. Another useful function of threshold archiving is the way in which this may be used to free additional space by deleting files for which current backups already exist. This action needs to be complemented by the ability dynamically to designate the files for automatic transparent restoration if and when they are called for.

When archiving involves tapes there is a danger that accidents will happen and that critical tapes will be overwritten. Tape management facilities protect tapes by recognising when they should not be overwritten. Several criteria can be used in order to decide whether or not to 'scratch' a tape. These criteria might be:cycles; generations; or time periods.

10.5 Aspects of security

Security is a very emotive issue in the world of business and commerce. It concerns corporate computing managers greatly that there is a strong access control in place. Early network systems were provided only with a rudimentary security system using weak password control. A good network security system should provide global enforcement of both user access controls and asset access. Asset is a term used to describe available applications packages or data. Asset and user access, if properly implemented, should allow access to be available only during specific hours on specific days by specific people. An important feature of security is the ability to record the attempts of unauthorised users to circumvent the system. Journal files can be used to discipline abusers within the company and can give valuable clues about current 'hacking' techniques, which can help in devising security strategies.

Some network operating systems require network administrators to set permissions for users and assets or files individually. Second generation tools allow a policy based definition, so that users may be granted facilities and services by virtue of their user name groups.

A famous hacking episode known as the 'Internet worm incident' occurred because users were allowed to change their passwords and then immediately change them back. Attempts like this and other cunning attempts at security violation must be detected and denied.

10.6 Problem management

System management needs quickly to identify the source and consequences of problems which arise. Such information is useful, for example, when system devices are rendered inoperable because of faults with parent devices. It is the nature of some problems that they are relatively benign until a particular function or device is called. It can be advantageous to 'track' progress and to profile faults due to software. Typically this enables software vendors' responsiveness to problem reporting to be observed by profiling problems reported and comparing profiles in later versions.

The better monitoring tools allow the observation by performance monitors of CPU; of processes; and of devices. Resource utilisation levels and bottlenecks may then be linked to any problems reported and tracked. A form of data logger might provide diagnostic evidence when matched to error report timings.

With accurate reporting facilities the system management can identify which processes and people are consuming computer resources.

10.7 Work load management

Casual users of networked computing call their applications and run them for their own purposes and in their own time. In a corporate environment there are often jobs that must be processed by the network but which have a more formal nature. In the past such jobs would have been executed by the corporate mainframe which was generally the preserve of the information services department. Whilst in more enlightened times each employee expects to have desktop access to significant computing power, the old jobs still have to be done. One example might well be the corporate payroll run but there are many others. These 'production' jobs are traditionally scheduled and run automatically on certain dates. These tasks are usually run conditionally on the execution of essential predecessors. For example, the payroll may run on the eighth of every month but only after overtime updates have been also run. Event based scheduling runs predetermined workloads when it is triggered by dynamic events. Such dynamic events might be a specific file being closed or a specific job having been started or terminated.

If a system management can call upon these facilities it is possible to plan the production schedule in terms of: sequencing; job and file dependencies; processing requirements; and business priorities. Some advanced management tools even allow the user to ask 'what if' for particular plans and simulate the effects of proposed schedules. This is very important when a network is shared between corporate production jobs and casual desktop users.

10.8 Printer spooling management

The everyday single user computer may well have an associated dedicated printer attached. With networking this makes little sense and the usual arrangement is to attach network printers so that all users can access printers. The problem created is the same problem that used to exist for multi-user systems. It is that it is necessary to move files for printing to the printer whilst maintaining their integrity and not jumbling them up. Complete print files are buffered and presented in an orderly queue. This function is often referred to as spooling.

10.9 Personnel related issues

A number of issues could be regarded as related to people. One issue is that of licensing control. Network managers must ensure that only authorised software is mounted on servers located in remote departments or made available through individuals' desktop machines. This is done to retain control of resource allocations and thus keep unpredictable software from causing downtime. To avoid appearing authoritarian, managers should be prepared to evaluate and approve if possible, software requested by users.

An often unappreciated task for network management is to provide training for potential users of applications. It is important that users feel comfortable with the network and more particularly that they have competence and confidence.

Earlier coverage of monitoring systems and devices implied that network administrators should act on exception reports or warnings that violations of predetermined thresholds. The keyword is 'act', since increasingly computer literate users will be quick to criticise centralised management control if performance falls below expectation in terms of speed, throughput and efficient archiving. To ensure adequate performance proper staffing levels must be acquired and retained. The levels may differ and are dependent on the maturity of the network in question. Typically accepted figures in corporate environments vary from one technician for every 50 workstations to one for a 100. Technician support is vital and should not be undersold for any short term justification of reduced costs. The figures quoted at the beginning of this chapter demonstrate the cost of downtime. The temptation is to reduce the staffing level when the network becomes stable but if the network is well configured and properly run the demand will naturally increase, allowing the retention of the initial number of staff but an increase in the workstation to technician ratio.

Over and above technicians, the network will need operations management especially in a corporate environment where the network has replaced a mainframe or minicomputer set up. Operations management staff will schedule tasks, deal with workload management, and monitor the system. Numbers should be chosen appropriately.

10.10 Architecturally related issues

The previous discussion was based on the client-server network model. Whilst the majority of local networks will operate on that paradigm, an increasing number of networks run distributed operating systems and there are also corporate long distance arrangements using data highways provided by commercial carriers. Distributed operating systems offer a network environment which includes servers but also adds the ability for an individual user to call upon extra CPU power. The extra power is derived from the unused CPU time of other users. Care must be taken in such architectures to allow for the possibility that no other users are switched on or that there is little CPU time free. The usual strategy for systems of this kind is the 'pool processor'. A typical arrangement is shown in Figure 10.2.

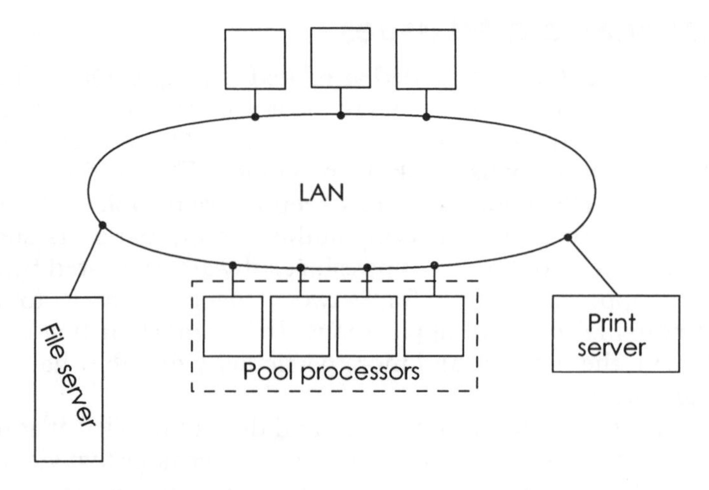

Figure 10.2 Pool processor architecture

10.10.1 Managing routing

Long distance networking using rented lines allows a company to design the interconnection of point to point switches. Although there are a number of theoretical models, such as the hierarchical network , in which routing is relatively simple most practical systems are examples of an irregular network as described in Section 5.7. In managing the routing tables in switching nodes is is essential to guard against two possibilities. These conditions are looping and oscillating. With the former, the network topology should be examined for links that allow a 'round trip return' or loop. A fragment of such a network is illustrated by Figure 10.3. To understand the difficulty, consider the following scenario: a packet is originated by node A and is destined for

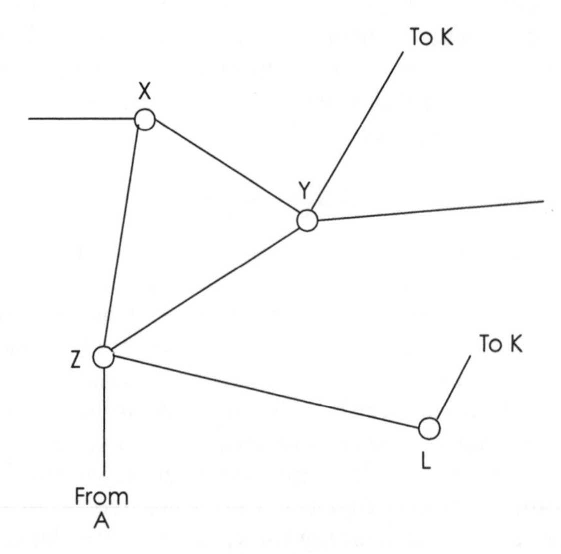

Figure 10.3 Looping

node K. The normal best route would be A,Z,X,Y,K. Owing to link failure, node Y is no longer connected directly to node K and has an altered routing table specifying that packets arriving at Y should be redirected to node L via node Z. Because of oversight in the case of manual routing or algorithm deficiency in an adaptive system, the routing table of node Z is not updated to take account of the new circumstances. Consequently, packets entering the system at Z are forwarded to X and then to Y, followed by a return to Z. The packet circulates Z,X,Y indefinitely. It never reaches K and merely adds to the traffic load on the links between X,Y and Z. The packet is said to be 'looping'.

The condition known as oscillating is simpler. Using Figure 10.3 once again consider a packet destined for K and originating at A. Assume this time that the optimum routing is directly from Z to Y and onward. Further assume that the link between Y and K is faulty and Y intends to redirect packets back to Z for transmission via L. Normally, the route from Z to K via L is an alternative but less than optimal. Consider the consequence of inconsistent table updating which has deleted the Z,L,K route. Z only knows one way to K which is via Y and thus a data packet will 'bounce' backwards and forwards indefinitely between Z and Y. This condition is called 'oscillating'.

If fixed routing is used, managers should establish procedures for node table updating which will not allow either looping or oscillating conditions to be created accidentally. In systems using adaptive routing network commissioning managers should ensure that the table updating algorithm effectively takes care of the potential problem.

10.10.2 Managing congestion

The underlying reasons for congestion are : links with insufficient traffic capacity to clear traffic building up at switching centres; and switching centres with insufficient buffer capacity to store traffic. Of course, the two reasons are linked but network management is responsible initially to specify adequately the capacity of links and switching centres. During the lifetime of a network the network management team is responsible for detecting potential congestion and installing more capacity or developing a policy to handle congestion. The traffic monitoring tools described in Section 10.3 may be seen to be vital in the early detection of incipient congestion. Once congestion exists in a system it grows naturally. This can be appreciated by thinking about the standard flow control mechanisms that might be part of the protocol. The simplest of these is the 'stop and wait' system described in Section 4.2.1. It will be remembered that an originating node sends a data packet and then waits for acknowledgement. If the receiving node is congested and has no free buffers it cannot accept the packet. The originator cannot obtain acknowledgement and thus cannot clear a buffer, and the problem starts to manifest itself at yet another node. The problem will tend to back propagate throughout the system.

Users are particularly irritated by the symptoms of congestion that are those associated with the non delivery of packets. There is a strong argument that suggests that users are much happier if they are told that no service is available rather than discover poor performance. This suggests that communication should use the virtual circuit or virtual call protocols first introduced in Section 1.3.5. By having a requirement to set up a virtual call the call request and call connect packets must traverse the system via the routing nodes. In so doing, buffer memory can be reserved or pre-allocated to a particular call. If insufficient buffers are available then the call cannot be connected and the user is so advised. Once the call is connected there is no danger of congestion because the available capacity has been allocated and overloads cannot occur. Congestion is much more of a problem when a datagram service is being offered.

One way of coping with datagram congestion is simply to forget any packets arriving at a congested node. ATM does this by virtue of its 'data loss' fallback strategy described in Section 5.6. A policy based on throwing away packets must be constrained by a sensible approach. This usually means that operations managers arrange for switching centre buffers to be subject to an allocation strategy so that a particular route cannot hog all the buffers, resulting in all other traffic having to be thrown away. It would not help the cause of congestion control if acknowledgment packets were cast away, since, dependent upon the protocol in use, the result will be a repeated data packet at some point. One often used strategy is to reserve a single buffer for packet 'examination'. Acknowledgement packets could then be recognised and not rejected.

Another alternative is to limit the amount of traffic allowed to be resident on the network at any time. Several suggestions have been made that involve 'permits' or 'tokens'. Nodes wishing to transmit a packet must first be in receipt of a 'transmit permit'. This gives rise to a big administration overhead and some difficulty in deciding how to control the permits. Yet another possibility is the notion that when congestion occurs on a given route, a control packet is returned to all nodes sending packets along that path. This control packet effectively says 'reduce your sending to node n by *x* per cent'. This last suggestion is very much based on a prevention philosophy of being better than a cure. The key to good control is the trigger used to initiate the control packets and several thresholds could be set in order to progressively choke off traffic for potentially congested routes.

It does not take much imagination to conceive the worst possible consequence of congestion. If communicating nodes have full buffers they can neither receive anything or transmit anything and a deadlock would exist. It is vitally important that network managers understand how their networks cope with congestion and that they have strategies for the recovery congestion, or worse, deadlock.

Questions

1 Discuss the issues inherent to network management and list for each issue raised the services required for effective control.

2 Describe the functionality of instrumentation necessary for the efficient monitoring of network activity. Enumerate the parameters that are vital to measuring performance.

3 Security is an essential element in network administration. Explain three examples of the implementation of security strategies.

4 Explain why spooling is necessary in networks offering a printer service. Include a brief definition of the spooling function and comment upon whether spooling is best carried out in the user's machine or the printer server.

5 What should be a network manager's attitude to computer virus infections? Suggest a strategy for dealing with the problem.

11

Distributed systems

11.1 Application layer issues

The case study organisation will need a powerful and large database. In the absence of a mainframe or single main computer a distributed or network system will be needed. Ideally each site will wish to store data predominantly used locally. Inevitably, other centres will wish to access data and there are a number of possible strategies for the running of co-operating databases resident on a number of machines on the network. The organisation may well need significant computing power which can be obtained by making the whole network of processors appear to be a single system. This implies that the local host computer is also able to distribute procedures among the other processors and remotely call those procedures when required. This arrangement produces the computing power by creating parallelism at the network level. Both the basic ideas behind distributed computing power and the use of that power in the context of a distributed database will be considered in more detail.

11.2 The distributed operating system

There have been many implementations of the distributed operating system. Each variant possesses strengths and weaknesses and focuses on slightly different objectives. There are underlying similarities which provide a basis for an overview, not least of which is the need for communication and the way it is carried out. The term 'distributed operating system' implies that every processor covered by the network is running the same operating system. Such a system is called an homogenous arrangement. It is possible for network processors to run different operating systems and yet be part of the same network. If this is the case there are a whole mass of different problems and each processor must be provided with an interface to provide translation. A process like this is often called an agent process and the complete network is said to be heterogeneous.

Care must be taken in using the terms homogenous and heterogeneous. There is at least one distributed operating system which may be run on heterogeneous hosts. As long as each processor is running its own native code version of the common operating system it is homogenous in that sense. The operating system is homogenous; only the processor types are heterogeneous.

11.2.1 Communications

The two main reasons for communicating are resource sharing and the calling of procedures which are resident on computers elsewhere in the network. From the point of view of the user the action should be transparent. Each machine in the network might have the kernel of the operating system running. The kernel of an operating system has traditionally been the part that deals with process scheduling, process switching, input/output operations and other basic and fundamental actions. In a distributed operating system the kernel also contains the communications subnet system, including issues associated with: the physical signalling; the data link layer; and the equivalent of the network layer. One distributed operating system relates to the layered model as shown in Figure 11.1.

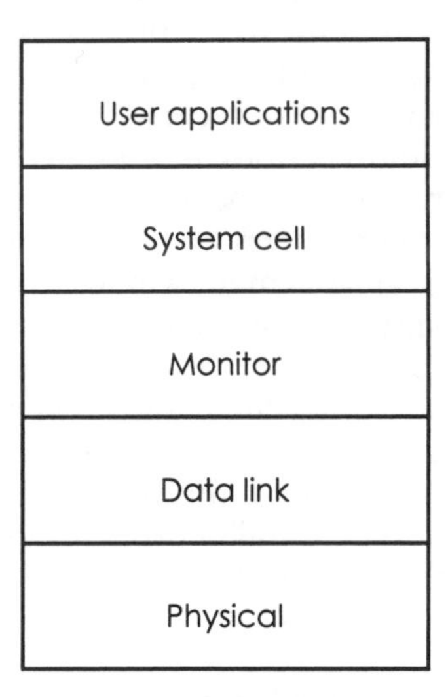

Figure 11.1 Distributed OS layer model

This operating system does not specifically define the physical and data link layers, allowing a number of options to be used. Typically the implementation is as an Ethernet or cheapernet broadcast bus with data packets delivered by a datagram style operation. The so called 'monitor' layer shown in Figure 11.1 is responsible for port addressing and mapping the port addresses to machine addresses. It also creates the data packets and hands them down to the data link layer. All of the hardware and software facilities are dubbed 'services', each of which is controlled by a particular process. Since these processes are concerned with controlling and sharing their associated resource they are referred to as 'server processes'. Each process is accessed through a port or ports which it has created and which it 'owns'. If the process resides in the user's local machine then the monitor layer of the system translates an port access request into the appropriate access addresses. If the process resides on a remote machine, then this will be evident by the port or port numbers involved and the monitor layer must direct an appropriately addressed request packet to a network interface. Port software must be able to handle queues of incoming requests and outgoing responses. Ports may also be used to allocate privileges or rights. When a process is initiated, it generates a port number and a set of 'rights'. A given port number might possess read only capability. A process may generate other ports with different rights as well. In this way, a process might have two ports, one read only, the other read/write. Because it owns two ports the process may receive data from either. Users, on being informed of the port numbers, may use the process in the way or ways defined by the rights allocated. For example, a process such as 'file service' may allow the reading and writing of files through port #271 and read only through port #5403. If as a user you have been told only of port #5403, then you may only read files. This is a powerful mechanism, but in a complex system a large number of port numbers will be required. In a well known system, when processes are created, a random number generator with a 48 bit word is used. Clearly, even in an ambitious setup,very few of the possible port numbers will be used.

Whilst considering communications it is as well to give thought to the nature of the messages that will be passed. At the simplest level existing processes on different host computers could be accessed by remote procedure calls (RPCs). Such an arrangement would invoke a remote process by naming that process and listing the parameters passed.The lower layers of the communications subnet would deliver the call to the remote machine which would execute the procedure using the parameters provided and return the result to the originating machine.There is more potential than this and many distributed operating systems do not limit the content of communications in this way. One, which supports a parallel programming environment allows high level program statements such as :

fork (process) on 2

The fork statement, when compiled, results in the process executable code being prepared so that at run-time it may be transferred over the network to network machine number 2. When the code has been downloaded the runtime system in the users machine will issue a remote procedure call causing process execution. If a repetitive calculation such as a specific matrix arithmetic problem or another algorithm susceptible to parallel computation is needed then a series of fork statements to a large number of destination machines could result in significant concurrent or parallel action. It is easy to see how parallel action would be useful to an organisation that needed regularly to evaluate field equations or deal with simulation or modelling in real time. The case study for this text is more of a straightforward trading business than a research and development environment and thus the majority of work done on the network will be different.The underlying work of a car breakdown and rescue service is very much one that will need a powerful database facility. Members' records would need to be consulted to ascertain the services they have subscribed to. Records of 'rescues' in hand, location of free breakdown trucks, and the status of current jobs will all be needed. From time to time it would make sense to analyse statistical reports in order to allocate or re-allocate resources for the most effective services and shortest call out delays.

The concept of an operating system that is operating on a network wide basis has given rise to the descriptive term 'middleware'. The implication is that this describes an entity which sits between the application and the physical network, provides remote procedure calls, context bridging , and the basis for management of a diverse resource rather than a centralised one. The problems of management have been described in Chapter 8. Middleware allows applications to task across the network system. Figure 11.2 illustrates the relationship between middleware, the network, application software and the operating system.

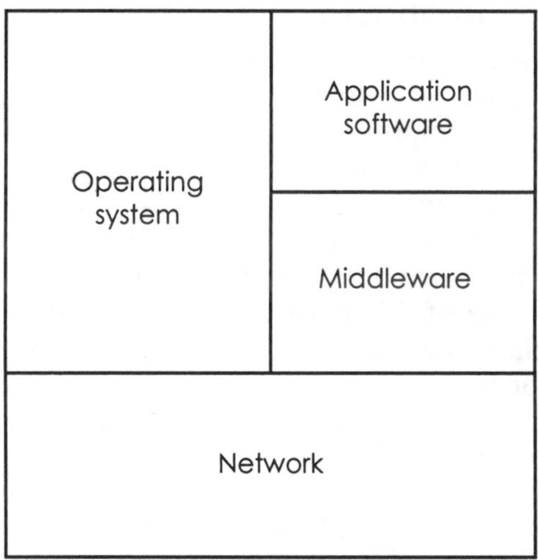

Figure 11.2 Middleware

11.3 Distributed data bases

Traditionally the magnitude of the problem presented by the case study would only be solvable by specifying a large mainframe computer. Each regional office would then have to access a central point for each database access event. To ensure reliable, any time access it would be necessary to support the system with a leased line or equivalent network with the inevitable associated expense. In addition, the system would be vulnerable to machine failure, which would affect all users and would be disastrous. One of the benefits of networked distributed systems is that it is possible to build in redundancy and duplication of services and information. Such a system is more robust than one that relies on a single central resource but is not without problems. To understand the nature of the problems it is necessary to review some basic database notions. Many database designs and architectures have been proposed and implemented but the universally most popular currently is the relational database.

Membership number	Name	City	Home region
059281	Hams, T	Linc	02
077246	Horace, C	Nor	04
100103	Haskins, J	Low	04

Figure 11.3 Member relation

11.3.1 Relational databases

A relational database is made up of a series of 'tables'. Each table or 'relation' relates certain attributes of database entries. Figures 11.3 – 11.6 illustrate relations which could be appropriate to the case study. It should be noted that one of the tables, often called the dictionary relation, is actually a relation of relations. Behind the concept of this tabular approach is the powerful justification that if everything is stored in tables then a whole unified set of management procedures for data manipulation becomes possible.

It is not the purpose of this book to describe in detail the management procedures but rather to investigate the problems presented to a relational database when it operates over a network. It is likely that the database for the case study organisation will be quizzed by users in such a way that the answers are available only by combining tables or relations in logical ways. Clearly, a query might arise such as, 'how many members have subscribed to the full range of services?'. To understand the detail of the necessary logical combinations you should look at texts covering query processing and query languages operating the principles of relational algebra together with relational calculus.

Service code	Services	Call outs this year
1022	Morning start & get you home	0
0032	Breakdown	6
1346	Breakdown & get you home	2
1223	Morning start	1

Figure 11.4 Services relation

Membership number	Service code	Price
059281	1022	70.50
077246	0032	35.00
100103	1346	85.00

Figure 11.5 Price relation

11.3.2 Justifying database distribution

The classic disadvantage of the centralised database is the potential vulnerability of the whole system to hardware failure or other cause. When a database is distributed, the situation is changed but just how is dependent on the nature of the distribution. In an example such as the case study it is likely that certain aspects of the database will often need to be accessed locally. Since vehicles spend, on average, more time within their home region then regional offices may justify the storage of member and vehicle data at the regional office. From time to time vehicles will be outside their region and a breakdown would result in the need for an access of one region's data by another, remote, region. This creates a number of difficulties. Indeed, one of the functions of distributed database design is to analyse the best distribution paradigm for a given case.

The splitting up or partitioning of the database forms the basic structure and affects the strategies needed for query processing or for database updating. To make a sensible decision it is necessary to take into account the expected number of transactions for each set of relations. For each case there

Relation	Location	Number of tuples
Member	Site 6	44095
Service	Site 2	40000
Price	Site 3	28000

Figure 11.6 Dictionary

will be a trade off between the limit situations of having complete copies of the full database at each location and having very little stored at a large number of locations. In the latter case there will be a high communications load and a relatively small local storage requirement, whilst in the former the opposite is true. The former case also has the disadvantage of a large updating problem when database elements are changed. Additionally, it might be said that there is a high cost for locations that do not generate much in the way of queries. The objective should be to optimise local storage and match communications throughput to available capacity.

To understand this point, consider a system which has a certain number of copies of subsets of the overall set of relations. More replications might be found in adjacent locations sharing relatively low performance data links.

11.3.3 Access response and updating records

If the distribution paradigm is good then the traffic caused by access requests should be minimised. When an enquiry is generated it is necessary to determine whether there is a local copy of the data required or whether it is available at some remote host. If it is available at more than one remote host, which is the best choice? The term 'best' might be defined in terms of least communication cost in time or money. One option might be closer but only accessible by a low bit rate link, another might be more distant but linked by means of and ATM or ISDN link. It is fair to ask first how the system 'knows' the location of different relation subsets. The relational database concept provides the answer. A relation or table can serve any purpose and each networked machine may have a directory relation in which the locations of desired relations are listed together with a 'figure of merit' defining the suitability of each possibility. The way that data is retrieved is affected by the details of how relations at different sites may ultimately be combined. It may prove advantageous to combine some relations at an intermediate point prior to the final destination.

The updating problem is truly significant in distributed systems. While a distributed system derives an advantage from the fact that data is held in diverse locations, there is also an implication that distribution allows concurrent or parallel processing. At any given moment several queries or updates to the database may be active. If two or more actions are demanded on the same data item there is a danger of inconsistent data. Some actions may read before an update, others after. Updates could occur in an order which leaves an incorrect final state. To appreciate this, consider a simple contrived example related to the case study organisation. To check that a member is entitled to a given service a regional office accesses the correct relation, which shows his current subscription is paid. Rather than just blindly add any new or additional subscription to the existing balance the accounting department normally reads the data, adds or subtracts services additionally requested and paid for or cancelled since last year. The update action is thus to read the data, modify it, and write it back. To add to the possible confusion, imagine that the client has normally only bought breakdown assistance. This year her spouse mistakenly renews this but adds car recovery. She spots this almost immediately and cancels the recovery' service. In the original database there is an original subscription for breakdown only. The accounts department will need to perform two database updates. The first increases the subscription, the second decreases it to its original value again. Assume that the original subscription had been £50 and that recovery costs an extra £25. The two transactions that accounting must perform, by sheer chance, are posted at roughly the same time. If reading and writing are allowed to be independent events then the following scenarios are all possible.

1. Accounts read £50; write back £75; read £75; write back £50.

This option is entirely benign provided that the regional office, checking on services subscribed to, does not access between the write back of £75 and the final correction.

2. Accounts read £50; accounts read £50; write back £25; write back £75.

In this case the cancellation is ahead of the request for recovery service. The client will be provided with a service they will have been refunded for.

3. Accounts read £50; accounts read £50; write back £75; write back £25.

Now the client will not be eligible for any services. Only option 1 eventually gives the right result. These issues are of even more vital importance to banks and building societies where the potential for chaos is clear. Problems like these did not occur when computers were not networked and machines were entirely serial in operation. The standard solution to the problem is to require a transaction to 'lock' a relation so that no other access can be made until the update is complete and the relation is 'unlocked'. This like enforcing a serial access, which begs the question 'how is the order of access determined?'. The simple option is to use a central lock controller to which all requests must go. The result would be a 'first come, first served' arrangement, but the traffic to and from the node running the controller would be large, possibly excessive. The system would also be vulnerable to failure of that one machine. Another option is to use timestamping. Each access request is 'stamped' with its time of origin, much in the way that files are often timestamped in order to sort or search only files generated since noon or some other time. This timestamping will have to have a finer resolution than the file example since requests for database accesses may be generated almost simultaneously.

Questions

1 Differentiate between the concept of a distributed operating system and a network operating system.

2 Explain why a parallel language requires an underlying operating system which includes message passing capabilities in its kernel.

3 Discuss the issue of crash recovery and fail safe operation in distributed systems.

4 Place the distributed operating system in the total context of parallel computing and justify the extra complexity by identifying applications that would benefit strongly from their use.

5 Explain vertical and horizontal partitioning in a distributed database.

Appendix A

Frequency changing

Carrier band and broadband networks rely for their operation on the ability to change frequencies from one value, generally the signalling baseband, to a higher range in order to move the signal away from undesirable interference. The process can be achieved using multiplicative or additive mixer systems. The additive mixer or frequency changer bases its action on a non linear device. Two signals, the original baseband and another called the carrier, are applied as input voltages to the nonlinear device. The simplest of electronic non linear devices is the semiconductor diode. Figure A1 shows the characteristics of such a diode. The graph plots forward voltage against the resulting forward current.

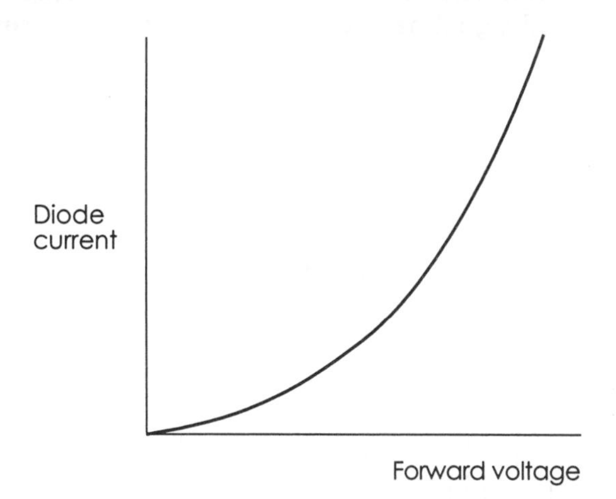

Figure A1 Diode characteristics

To understand the action, consider two sinusoidal signals, sin A and sin B. Sin A might be a carrier in the order of 400 MHz. Sin B might represent the baseband fundamental signalling component, say 5 MHz. The characteristic equation relating output current (y) to input voltage (x) is:

$$y = ax^2 + bx + c$$

In this case c = 0 and therefore it can be said that $y \propto x^2 + x$.

Since

$$x = \sin A + \sin B$$

then

$$y \propto \sin^2 A + 2\sin A \sin B + \sin^2 B + \sin A + \sin B$$

Neglecting all components present in the output except $2\sin A \sin B$, since they may be filtered out, it is appropriate to recall the trigonometric identities for $\cos (A - B)$ and $\cos (A + B)$:

$$\cos (A - B) = \cos A \cos B + \sin A \sin B$$

$$\cos (A + B) = \cos A \cos B - \sin A \sin B$$

Thus

$$\cos (A - B) - \cos (A + B) = 2\sin A \sin B$$

This shows that in the output there are sum and difference frequencies of the carrier $\sin A$ and the baseband signal $\sin B$. The fact that the sum and difference components are cosine waves just means a 90 degree phase difference from the sine inputs and this is insignificant. Taking the original values for $\sin A$ and $\sin B$, the output would contain a signal at 405 MHz (and one at 395 MHz). The same information that was present on the baseband 5 MHz signal may be derived from the 405 MHz signal by using the original 400 MHz carrier as a reference against which to demodulate. Figure A2 shows how the baseband signal has been moved up the spectrum to clear noise and interference.

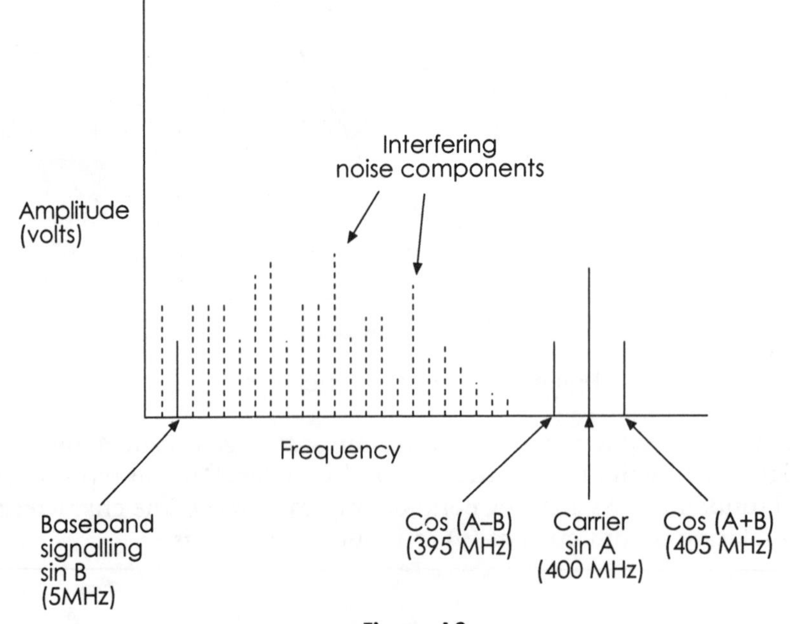

Figure A2

Appendix B

Filtering

In computer networks it is often necessary to select or reject signals according to their frequency. Selecting sum or difference frequencies in carrier-band networks whilst rejecting noise or interference is one example; selecting a particular channel in frequency division multiplexing or broadband systems is yet another. Electronic circuits may be devised as the filters. These circuits are classified according to the range of frequencies that they accept or reject. If they accept a range they are said to 'pass' that band. If they reject they 'stop' a band. The complete set of possible designs is:

1. Low pass – which passes all frequencies below a given value.
2. High pass – which passes all frequencies above a given value.
3. Band pass – which passes frequencies between a lower and upper frequency limit.
4. Band stop – which discriminates against frequencies within lower and upper limits.

Figures B1 (a) to (d) illustrate the ideal characteristics for each of the options. Ideal characteristics are often called 'brick wall' in nature since they have sharp and absolute differentiation between frequencies they pass and those they do not. The reality is that the ideal cannot be achieved and filters have a 'roll off' rate. Even so, there are 'cut off' frequencies denoted f_c associated which each filter. The quality of a filter is assessed in terms of two factors: the rate of roll off; and the gain at the cut off frequency. Figure B2 shows a more realistic characteristic for a low pass system. Gain and attenuation in filters and electronics generally is often expressed in decibels. The decibel is a logarithmic measure which for voltage levels is calculated from:

$$20 \log_{10} \frac{\text{output voltage}}{\text{input voltage}} = \text{decibels}$$

The values of gain or attenuation for a range of voltage ratios are given in the Table on the next page.

Voltage ratio output/input or (input/output)	Gain or (attenuation) in decibels
2	6
5	about 14
10	20
20	26

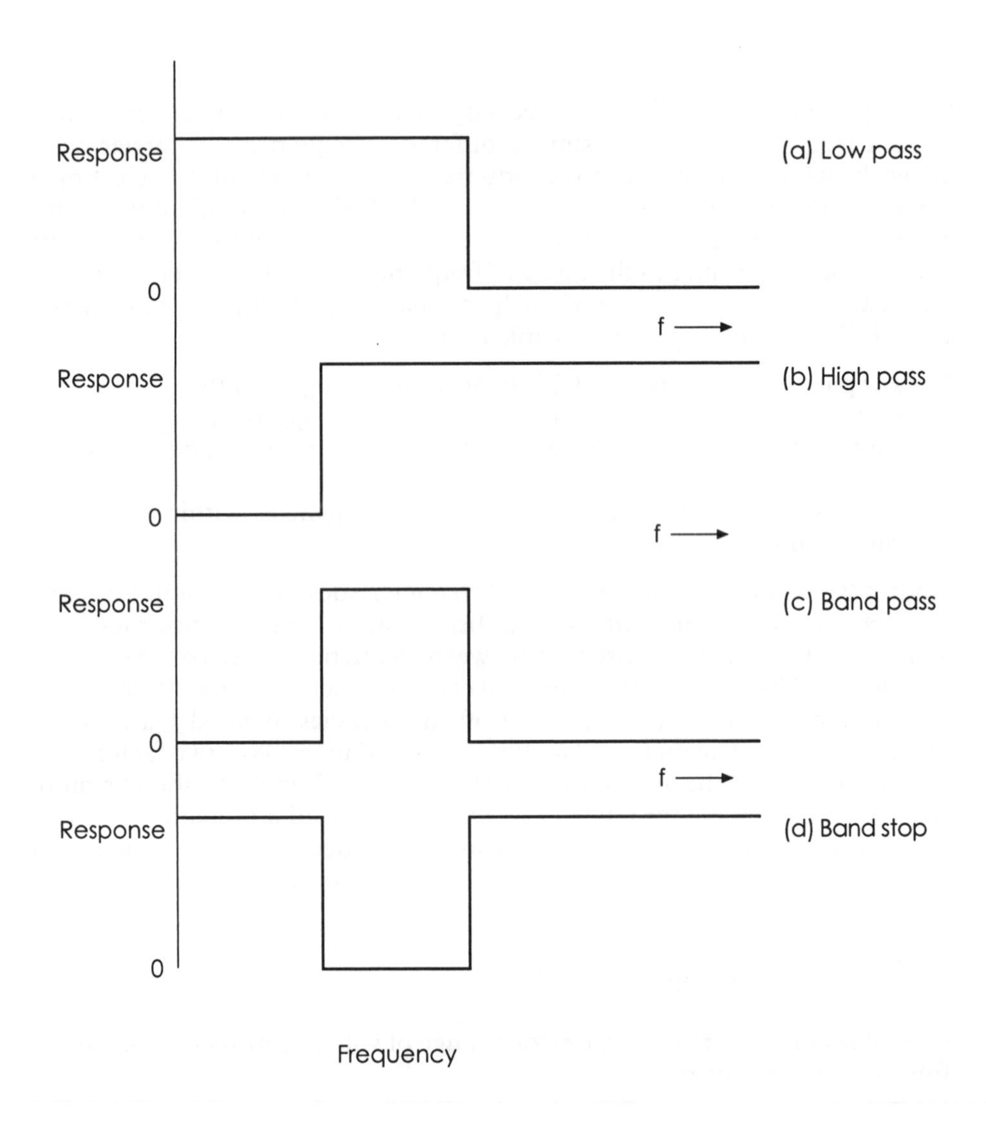

Figure B1 Ideal characteristics for filters

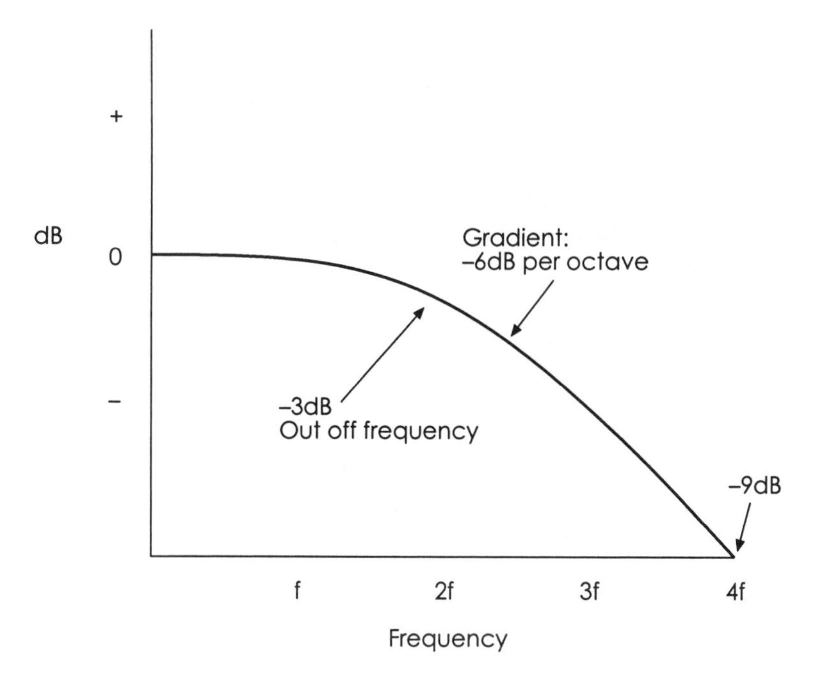

Figure B2 Realistic characteristic for a low pass system

The cut off frequency for a filter is defined as the frequency where the voltage output has fallen to 0.707 of what it was in the passband. This corresponds to –3 dB (decibels). In talking of frequency ranges two common values are : the octave – an octave is a doubling of frequency; and the decade – a range of 10:1. Filter roll offs are naturally expressed as a certain number of decibels per octave or per decade. It is very easy to make a filter with a roll off of 6 dB/ octave which corresponds to 20 dB/decade. Unfortunately, this is far from sufficiently adequate for practical applications which might require better than 80 dB/decade. It will now be evident why 'guard bands' are needed in broadband and in radio channel arrangements. They are necessary to allow filter roll offs to 'get between' wanted and unwanted signals.

Appendix C

The ISO OSI 7 layer model

In Chapter 1 the model was introduced but the jargon associated with the model was minimised. The standard names for the layers are given below together with a brief rationale for each.

Layer 1 The physical layer

In Chapter 1 this was called the lowest layer. It is the level at which electrical signals are exchanged. Design of the layer is concerned with issues such as definitions of voltage levels and tolerances acceptable for 0s and 1s. All the physical connections must be defined. Examples are: the media specification; and plugs and sockets and their pin standards. The design issues are mechanical and electrical.

Layer 2 The link layer

The provision of an error free link is the aim. The design issues of this layer are all about the data frame operation. Error detection, correction strategies, acknowledgment, and flow control for the frames are also part of the requirement. The layer name is derived from the fact that a basic data frame operation provides a data link.

Layer 3 Network layer

This is the data packet operation. Packets are necessary when data is to be exchanged over different segments of a large network and where routing or re-routing become necessary. While the data link layer only provides local net services such as acknowledgement, the packet layer offers end to end services. The key design issue concerns how routing is determined and how any possible congestion might be handled.

Communication subnet

The term communication subnet is often applied to the first three layers. A basic communication service with no frills except a reliable connection could be provided by just layers 1, 2 and 3. Because of this organisations which carry data for reward may present the customer with a communication subnet service. It is up to the customer to superimpose any further layer functionality needed for his own business purposes.

Layer 4 Transport layer

As described in Chapter 1 the main interest at layer 4 is the end to end control of the dialogue. In more detail, 'control' means: breaking data down into appropriate size ; multiplexing a number of channels of data; sequence control and ordering or re-ordering of data blocks that may have been passed over variable delay routes; issuing commands down to the network layer to establish, use or clear packet level communications. Congestion is dealt with at layer 3 in the sense that layer 3 ensures that nodes are not 'overpowered' and backed up by traffic offered. The transport layer has a related role but is concerned with ensuring that a particular host is not overpowered. It provides a regulatory function when a fast host is connected to a relatively slow one. For this reason layer 4 is often referred to as a host–host layer.

Layer 5 Session layer

When a network user wishes to speak to another user, a 'session' of communication is needed. Software from layer 5 has to check that such a communication is allowed or authorised. Another task is the negotiation of the details of the session by agreeing mutually acceptable protocols like half or full duplex working. Once a session is under way the layer 5 software is responsible for injecting 'way points' so that, if communication is lost, an orderly recovery is possible. The effective operation is one of 'binding' the users together.

Layer 6 Presentation layer

Layer 6 provides high level data processing. It may map various possible data formats into a form suited to the nework, or suited to the designs of the users. Typical compression and encryption schemes appropriate here have been fully described in Chapter 8.

Layer 7 Application layer

The job of the software at this level is to provide hooks to allow interchange of data between different users' application programs. The application layer is the boundary between the network and the users' programs. Typical user programs at this level are:file transfer; distributed data bases; e-mail; and distributed operating systems.

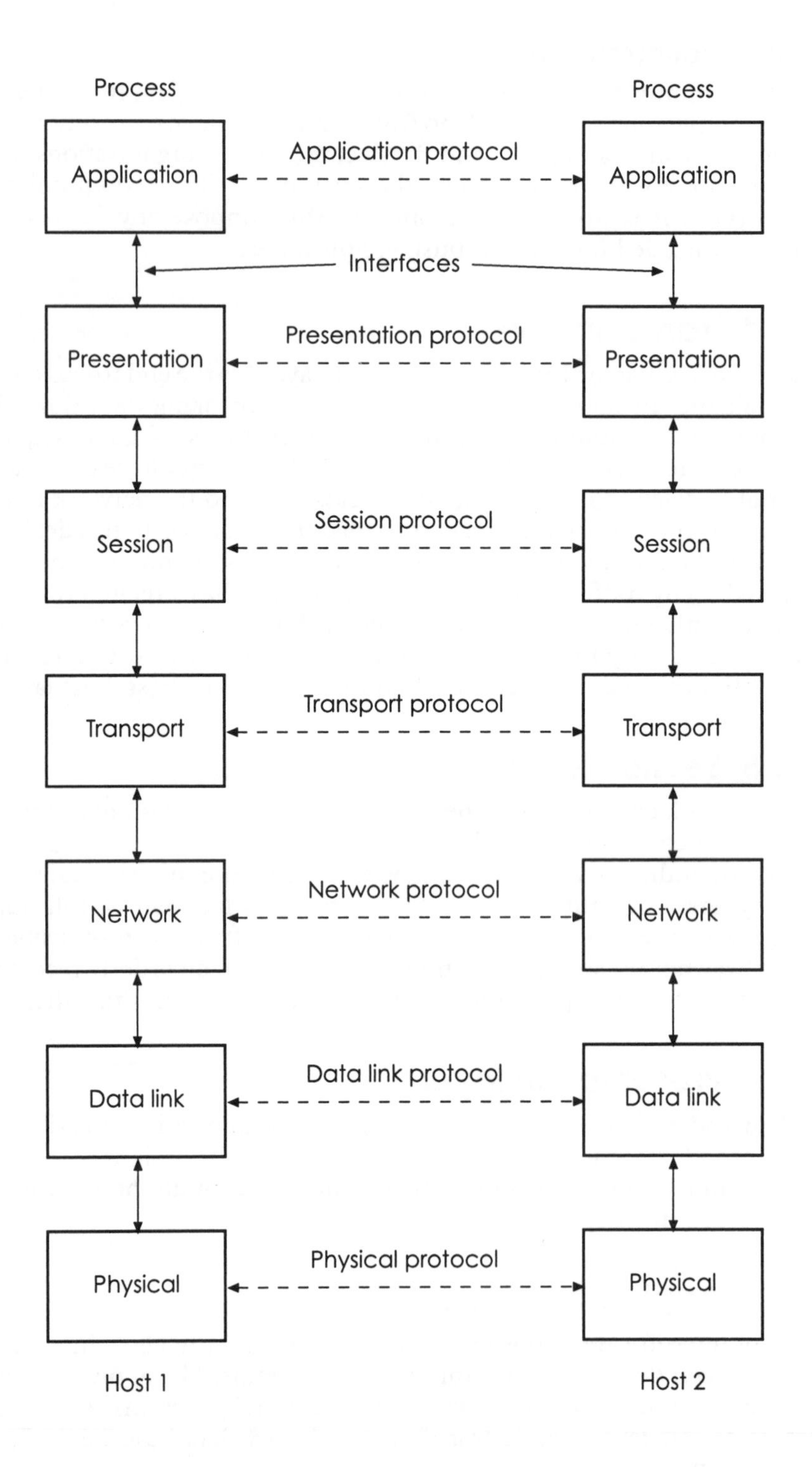

Figure C1 The ISO 7 layer model

Although the layers have been presented as the defined entities it is equally important to realise that to achieve standardisation without stunting the development of the layer function, the interfaces between the layer processes must be rigidly defined. Part of the definition of a given layer is devoted to how it interfaces upward to the layer above and downward to the layer below. Good interface definition allows internal developments within the layers to evolve productively.

It will have been noted that in some layer descriptions the term 'the layer software' has been used. Generally, layers 1, 2 and 3 may all be implemented as hardware. They need not be, but a hardware version is entirely possible. Above the subnet the implementation of the layers is exclusively as software.

A false impression might be gained that a network that follows the model must implement all of the layers. This is not so. There are many cases of networks designed for manufacturing or process control in which a large number of the upper layers have been omitted. They are often omitted to improve the speed of the network, host to host.

Another piece of jargon associated with the model is the notion of peer to peer correspondence. The notion is rather abstract, in that it imagines that if an 'observer' were to 'drop in' at any level on both sides of a link he would see the same activities going on using data of the same type. Another way of thinking of this is to imagine that in Figure C1 on the previous page everything below a certain layer was 'chopped off' and that there was a link established between the two ends at the lowest remaining level. The 'removed' layers are transparent and it is as if communication may take place between any two peer level ends, i.e. between two session layers or between two network layers, or transport layers. It is abstract of course, because actual communication only takes place at the physical layer, where there is media actually connecting the ends.

Among the criteria reputed to have been used by ISO in their development of the model are the following:

- Layers should perform a well defined function;
- There must be enough layers to avoid combining unnatural partner processes at a given level;
- Layer boundaries must be designed for minimum data flow between layers.

Appendix D

Answers

Chapter 1 The requirement

1. The design and implementation of a data communications network is a non trivial task. Inevitably, it would involve a team of workers. A developer might choose the 7 layer model because it allows a top down design approach in that the overall problem may be decomposed into a set of subsystems. A sub team of appropriately skilled personnel could work on each of the subsystems. The project manager must ensure that the interfaces between the subsystems are well understood and defined. The beauty of the 7 layer model is that it defines, in a standard way, the functionality of each subsystem and the messages that must pass between them. The skills needed by the personnel dealing with layer 1 are generally those of the electronics engineer. Layers 2 and 3 may well have all or part of their function implemented in hardware but in these cases the hardware engineering skills must be enhanced by those with a knowledge of data network access methods and all aspects of protocols. For the higher layers the implementation is probably all software and thus software engineering skills are necessary but these must, in turn, be enhanced by specialist knowledge of communications data processing and the interfacing of applications packages to the network.

2. The developer would normally be keen on a layered model because it would enable the individual development and upgrading of given layers so that the product is improved incrementally. In this way an improvement to an existing network could be tested adequately before release. Another big advantage to the developer is the opportunity to enter into agreements with other vendors. A particular network might have been designed to operate with a specific lower layer protocol and because of this the market may not be fully open to the product. An arrangement to adapt the interfaces to suit another vendor's lower layers might open up further markets. If layers are amenable to individual development then the higher layers may remain unaltered.

3. The peer to peer correspondence in layered models refers to the fact that data entering the network at the application layer is transformed and

operated upon by each layer. Communication between the layers is carried out in a standard way and thus if two ends of a link are considered and furthermore, if each end is decomposed into the hierarchical layers, then it should be possible to imagine 'dropping in' at the interface between any two layers. If this is done at both ends during a communication the observer would 'see' the same data in the same format. Because of this it is suggested that it would be possible to imagine that it is as if an end to end link exists at any layer interface. In truth these links are purely virtual and the only physical link is at layer 1 but the concept gives rise to peer (or equal) layers and a correspondence between them.

4. A protocol is a set of rules for a conversation and as such it should address the following:
(a) A format for exchanged entities such as data frames and packets.
(b) Details of how they should be exchanged including arrangements for acknowledgement.
(c) A specification for an error control coding option, either error detecting or error correcting.
(d) Details of the arrangements for crash recovery and reset.
Clearly there is likely to be partitioning of the overall protocol, with specifications being given for higher and lower levels.

5. The EIA layer 1 standards are concerned with the electrical signalling between nodes or stations. They do not specify whether the transmission is synchronous or asynchronous but they do concern themselves with: signalling voltage levels; a definition of what each circuit should do, i.e. transmit,receive data, ground, handshaking and modem control circuits; a definition of the plugs and sockets to be used; maximum cable lengths; and the specification of the sourcing capability of line drivers used and the sensitivity of receiver circuits with their consequent effects on maximum signalling speeds and range. A number of standards exist, including one designed for operation in harsh electromagnetic environments where interference is highly likely and of a significant level. For this, the appropriate EIA standard calls for a balanced signalling system with each link direction being two wires driven in antiphase against a signal ground reference. The receiver is of the differential type and interference induced into the lines will be cancelled.

Chapter 2 Data transmission

1. Digital modulation is the process of changing a digitally coded signal into another digital encoding. It is thus a question of format and how a fundamental '0' or '1' is actually signalled. The standard format where a '1' is signalled as a positive level and a '0' is signalled as a near reference ground (0 *v*) is the best known but it is unsuitable for many applications, and alternatives like Manchester encoding (Biphase) or WAL2 may be specified by communications link designers.

2. Formats other than the standard offer significant advantages when transmission is synchronous. For synchronous systems the receiver relies on being able to identify accurately the sampling instant for each bit in a possibly continuous stream of data. If an appropriate format is chosen it will be possible to recover the original clocking information from the signal itself and consequently determine sampling instants. Appropriate formats are guaranteed to contain harmonics which when filtered from the signal allow this to.be done. To understand, consider alternate zero crossings of the second harmonic of a basic standard data stream. The problem is that for the permanent presence of the second harmonic the data stream must not be 01010101010101010. Obviously, in real streams other patterns will exist most of the time. The purpose of digital modulation is to always provide the correct harmonic content at all times irrespective of the patterns being signalled e.g. 11100000111001010110011111111.

3. In asynchronous signalling the link is genuinely idle when there is no traffic passing. Receivers are woken up by the first transition from the idle state (start bit) and synchronisation of receiver sampling is achieved by counters driven by nominally similar oscillators at transmit and receive ends. If more than a few bits were signalled without re-synchronisation then the difference between the oscillators at the two ends will cause a gradual drifting out of lock. To guarantee a 're-synch' on a regular basis there is an enforced idle period after each byte transmitted (stop bits). Each new byte is preceded by a new start bit. Synchronous systems do not rely on counters and start/stop bits. Instead they either have separate clock circuits or they use clock extraction as described in Answer 2 above. Synchronous systems are either permanently active, sending signals which say 'I have nothing to send'; continuously during idle periods; or they attach 'run in' sequences to the front of any communication to allow receivers to lock into bit synchronisation before any data of consequence is passed. They may also use word synchronising techniques when they are bit oriented. The specific cases are best served as follows:

(i) The high traffic density of the first case implies that the overhead that must be paid for each asynchronously transmitted byte, i.e. start and stop bits, would significantly use link bandwidth and thus reduce capacity. It would be sensible to use a synchronous transmission system.

(ii) The second option defines a link that would generate only very sparse data. Even the best of typists can generate only 10 characters a second. This corresponds to about 80 bits per second. The screen of a dumb terminal is adequately refreshed at 1200 bits per second, corresponding to more than a line per second. Most readers cannot read much faster than this. The maximum rate is thus only about 120 characters per second or one approximately every 8 milliseconds. Such a rate does not pressurise the meanest of serial links and so an asynchronous transmission is adequate.

4. An industrial environment is full of machines which generate interference. This may be caused by high speed transitions in inductive machines, sparking commutators, and supplies to large motors being carried in adjacent cable trunking to that used for networks. The frequencies created by such interference extend from a few hertz to about 400 MHz. A baseband network would share the spectrum with this 'noise', and data corruption would be perhaps unacceptably high. The use of analogue modulation techniques allows the data signals to be superimposed onto a carrier above 400 MHz and outside the main interference zone. This would strongly improve the error rate for a system operating in an electromagnetically harsh environment.

5. Guardbands are regions of the electromagnetic spectrum which act as separators between signals. They are created by allocating carriers frequencies sufficiently far apart that filters of a reasonable order can discriminate between signals. The worst case bandwidth of signals in a broadband system must be known, as must the skirt characteristics of any filters used before appropriate guardbands can be determined. Althought the question considers guardbands in the context of broadband systems it will be recognised that a similar requirement exists for radio networks using multiple channels either for data or for control purposes.

Chapter 3 Single network issues

1. In a ring structure the main data highway forms a closed loop, whilst a bus is a highway terminated at both ends in an impedence matching the characteristic impedence of the line. Generally, nodes for the two options are slightly different in operation. Bus nodes broadcast directly and all nodes hear the transmission. Rings tend to have nodes which 'break' the highway when they wish to transmit.

   A token bus simply has a special data frame which circulates around all the nodes on the system. Each node knows its successor and addressess the special token frame to that successor. In this way, although the token bus is physically a bus it operates as a logical ring. Clearly, the network must sense if a node is not on line and thus deal it out of the logical sequence for token passing. Similarly, it will need to recover if a particular node goes faulty. Such arrangements have been devised in a number of designs, including MAP, the manufacturing automation protocol, which needs the determinism provided by token operation as well as the convenience of bus distribution.

   A contention ring expects each sending node to transmit its data frame around the ring and receive it again after a circuit. Destination nodes read the data frame as it passes their node end. Instead of the orderly wait for a token, contention rings simply allow their nodes to transmit when they wish. If two transmit together a data clash is registered because they check that their returned frames are the same as those they sent and if two nodes are transmitting they each receive the other's frame. On detecting a clash they fall back, and wait a random time before trying again.

There is no rigid link between access method and topology as has been demonstrated.

2. The truth of the statement depends upon what is meant by 'orderly rules'. Because of the way in which the question is framed it is assumed that straight contention is considered disorderly. Orderly systems involve access methods such as polling or tokens. In these systems a node may not transmit without either having been asked explicitly if it has any traffic for the network or alternatively if it is possession of a permission to transmit in the form of the token. A straight contention system simply allows its nodes to transmit when they wish and this is bound to cause a lot of data frame corruption, since there would be many instances where frames would be almost finished or part way through when they were interfered with. For a case like this it would be impossible to quote a worst case delivery time for a data frame. An attempt to send it might take place just once, with success, or one, two or many times. Conceivably a straight contention system could become jammed solid in the presence of very high traffic offered. Contention may thus be classified as a probabalistic system. CSMA/CD does introduce some rules such as 'listen first', but data frame collisions may still occur when nodes transmit within the signal travel time on the highway. For this reason CSMA/CD is also strictly probabalistic. If a delivery time (worst case) can be guaranteed, as in token or some slotted systems then the classification is deterministic.

3. Synchronous data link control was devised as a 'terminal cluster' network connecting terminal servers to terminals as part of a larger network where the backbone highway operated at much higher speeds. SDLC is adequate for interconnecting and networking small computers and relies on the polling access method. One station acts as the master node and regularly issues poll requests to each slave node. The sequence can be straightforward hub polling, e.g. 1,2,3,4,1,2,3,4,1,2,3,4 etc. The master acts as an intermediary in all communications so that if node 3 wished to send a frame to node 4 then node 3 must wait until quizzed by the master, forward the frame in response, the master 'selects' 4 and forwards the frame. The highway is a bus structure. The protocol used allows multiple frames to be sent before acknowledgment and in doing so indicates the next frame expected. In this way, if five frames were transmitted and number 3 was corrupted, the acknowledgment would specify 3. This implicitly acknowledges 1 and 2 but causes a repeat of 3, 4 and 5. Node control, i.e. switching on and off, resetting count number for frames and restarting in the event of problems is available to the master. For this reason there are data frames, and control frames. Some control frames are unnumbered. Each frame is bit oriented and has an identical flag for opening and closing frames. This flag is unique because bit stuffing precludes it from appearing within the data fields of frames.

4. Normally CSMA/CD nodes listen first before transmitting. This avoids most potential data collisions but cannot avoid the circumstance that arises when two nodes transmit almost simultaneously. The signal travel time of the line means that, although a node is already transmitting, the signal may not have reached another yet. After listening, it erroneously concludes that no-one else is transmitting and does so itself. CSMA/CD requires nodes to continue listening to their own transmissions and in this way a collision is eventually detected. No two frames can be the same since at minimum their source addresses are different and thus also their checksums. At some point one node will be asserting 0 on the highway when the other is asserting 1 and the clash is detected. CSMA/CD requires that in the event of such a clash detection, both nodes stop transmitting and wait. Each contains a random number which is used as a delay counter and they try again to send their respective frames after their individual time outs expire. Data transmission tends to be bursty and this strategy should iron out the trouble but it is possible that a node, on trying again, clashes with a different transmission. In this case the delay is increased by a binary weighting, e.g. 2,4,8 etc. before another try occurs. The role of the binary back off is to spread traffic offered peaks.

5. Long or short data elements are assumed to be erroneous. Most protocols will define maximum and minimum frame and packet sizes for their system. Short frames are often the residue of a data collision. As such they are ignored and just represent wasted bandwidth. Long elements are generally detected as misaligned frames since extra bits exist in one or more of the data fields. A misaligned frame will unquestionably be found to have an incorrect checksum or CRC and therefore will not be acknowledged. Such frames will normally be repeated in most protocols.

Chapter 4 Error free channels

1. The merits of the three error detection schemes are directly proportional to the effort required in generating them. Parity is almost trivial to implement but is the most vulnerable. Block checksum is more difficult but improves the performance, whilst CRC is really very good but has the most difficult algorithm. Parity bits have been most commonly used in byte oriented asynchronous serial communication. Very simple hardware in the form of cascaded XOR circuits can produce the required parity bit and this simplicity has allowed asynchronous serial ports to provide automatic generation and checking of parity. Block checksum is still a parity system but in addition to producing a word parity in the normal way it treats a group of words as a block and creates a new word comprising the vertical parity of each columnar bit in the group. It gives a two dimensional parity check which leads to improved performance. CRC or cyclic redundancy check is suitable for bit oriented work, as a CRC word can be generated for a word of any size. Because it provides a 'signature' for a bit

stream it encapsulates not only the 0–1 transitions in terms of their correct number but also in terms of their exact position. The feedback mechanism used in a CRC generator exagerrates any differences so that a single bit error causes a large change in signature. The benefits of CRC more than outweigh the more complex generation and checking process because it is strongly sensitive to burst errors, which are very common in networking.

2. The CRC is basically a modulo 2 division of a bit stream by an agreed polynomial. Once the division has been completed there will be a remainder (even if it is 0). The remainder is appended to the message word from which it was generated and when the receiver subtracts the remainder and divides this new word by the agreed polynomial the result should have no remainder. If it does then the word has been corrupted in transmission. The given polynomial is 10101 and the message is 11100110. Dividing the one by the other creates a remainder of 10100. The remainder is subtracted modulo 2 (XOR) from the original message word which is now divisible by the generator polynomial. In this case the process yields 11110010, which when divided by 10101 gives 0. This illustrates the principle but not how the CRC is appended. If the message has appended to it the number of zeros equivalent to the number of bits in the generator the message treated as a number will be multiplied by 2,4,8,16 etc. it will still be divisible by the generator polynomial (appropriately scaled), with the same remainder. If this remainder is subtracted (modulo 2, XOR) it will affect only the appended 0 bits of the message and because in modulo 2 addition and subtraction are the same, a message such as 0101110000, which when divided by the generator polynomial gives a remainder of 101, would become 0101110101. Now, the remainder is seen to be appended to the back end of the message.

3. A go-back *n* protocol allows frames to be sent in multi frame groups but should an error occur in transmission then acknowledgment is only given up as far as was good in the group. Say a group of frames numbered 4 to 9 were transmitted. If number 6 was corrupted the receiving end would acknowledge up to number 5 and the transmitter must send numbers 5, 6, 7, 8, 9 again. In other words it must go back *n* and repeat the rest. A sliding window protocol has a window of expectation for the numbers of any frames received. In the case of the question, the window size is two and thus if frame 5 was the last acknowledged then only frame 6 and frame 7 are legitimate new receipts. In most implementations the number is greater than two but even this allows for frames to be received out of order. The job of the sliding window node is to re-order out of order frames and present them to the host computer in the correct numerical order. In many cases frames received outside the window are discarded without comment. Should a number of out of window frames be received the sending and receiving nodes must resynchronise their models of the frame number situation. Non acknowledgment usually results in the sender sending a duplicate frame and in the case described, only frames 4 and 5

may be acknowledged. Then 6 and 7, then 8 and 9. The sequence might be: 4,5,6,7,8,9; 6,7,8,9; 8,9. The detail depends very much on the specific action of the protocol and variations abound.

4. Character parity in asynchronous transmission will only detect errors that occur in odd numbers of bits e.g. 1,3,5,7 etc. It is vulnerable to compensating errors occurring in even numbered pairs, so it will be helpless if an intended 1 is received as 0 and in the same word a 0 is received as 1. The relative power of this system is very low. It is a token gesture but, as with many things, more than double the effort results in less than double the performance. There is a law of diminishing returns for effort in error detection. Block check methods will detect all but 2 bit errors occurring in the same column in the same block. Cyclic redundancy check is particularly good at detecting burst errors. These errors are very common in data communications with whole groups of bits being wiped out. In general the method will detect all of the burst errors affecting up to the CRC word length and a very high proportion of longer bursts.

5. Block systematic codes are implemented by processing blocks of data at a time. This means that a message is broken down into short blocks and then the necessary additional bits are added to produce the full code word. The number of additional bits is prescribed by the error correcting performance needed. Convolutional codes expect data to be provided in a continuous stream and input into a shift register which has a series of half adders monitoring certain bits and generating extra bits from the combination of bits currently in the register. Block codes, however, tend to have a neat split between message bits and error correction parity bits convolutional coders have an output that is much more of an integral nature.

Chapter 5 Interconnecting networks

1. The key in differentiating between repeaters, bridges and gateways is the protocol or protocols operating on each side of the device. If the two sides are operating the same protocol and the link between them is essentially for the purpose of extending the net, then the device is a repeater. Dissimilar network operating conditions may be connected by means of a bridge. Bridges must interconnect networks under unified adminstration, even though they might have different operating conditions. The bridge may be viewed as a device which translates at the level of the data link layer. The gateway is a device which is able to interconnect networks that are completely dissimilar. The administrations may be independent because a gateway can operate by translating at the highest levels. The lowest translation to be expected would be the network layer and the system might even operate at the application layer. A multiport repeater is a convenience in structuring network segment installation. The case study head office building might benefit from a centralisation of the segment joining repeaters.

2. The fibre distributed data interface (FDDI) was designed to cope with increasing demands for longer range for backbone networks. The use of optical fibre also provides a much greater bandwidth and thus the ability to provide higher data transmission rates. Apart from the extension requirement for ordinary commercial data networks, variations have been the use of FDDI principles in military aircraft avionics. FDDI nodes may be class A or class B. Class A nodes are connecting to both primary and secondary rings. Class B nodes are connected directly to the primary ring and they are not connected at all to the secondary ring. Primary ring connections are made through a special wiring concentrator.

3. The scope of X.25 is limited to the connection of users to the nearest digital exchange of the data carrier. The PAD at the user's end delimits one end whilst data transmission over the carrier's network operates under a completely different set of protocols and thus an PAD at the exchange end marks the other limit. It is the carrier's business to ensure that the users packet data reaches the intended remote destination where, at the nearest digital exchange, the data is repackaged as X.25 for onward transmission over a local digital line to the destination PAD. The difference between X.25 and Internet is reflected in their respective origins. The Internet has its origins in the US Department of Defense and its requirement for a network routing protocol. This provenance dictated a system based upon the datagram. X.25 was originated by the CCITT and is prescribed by the preferences of the telecommunications industry. The history of telecommunications has prioritised the concepts of circuit switching and the 'call', which has set up, conversational and clear down phases. As might be expected, the CCITT (the International Telegraph and Telephone Consultative Committee), has formulated X.25 on the basis of the virtual call. In 1980 the standard was modified to include a datagram option but this was removed at the next four yearly review of the standard which occurred in 1984. No commercial network ever offered the service.

4. In asynchronous transfer mode (ATM) connects high speed data highways by means of crossover switches. These switches must cope with the fact that much data traffic is of a bursty character and this, combined with the possible delays in waiting for output multiplexing slots mean that 'in line' buffers and their management assume a high level of importance. Every effort would be made to provide sufficient buffering to cope with all expected traffic level contingencies but inevitably the system may become 'embarrassed' on occasions. ATM solves the problem by discarding data cells. There is an agreed maximum discard rate for each system but ATM must be regarded as lossy. The expectation is that users will provide error control and recovery at a higher level if the data is sensitive to loss.

5. A review of the general layout of the case study network reveals a hierarchical structure with relatively few regional centres linked by ISDN, X.25 or private lines. It is true that the local area networks at each site are

relatively complex but the intersite links are simple. In many cases there are no alternative intersite routings possible. Adaptive routing is more applicable to complex interlinks where alternatives do exist. The conclusion is that directory routing is more appropriate. The consequence of this choice is that the network management must be well informed about network traffic statistics for each of the links.

Chapter 6 Going further

1. The characteristics of the public switched telephone network (PSTN) show that there is a low and a high frequency roll off, The lack of response below 300 Hz may be attributed to transformers such as the two to four wire hybrid type. At the high frequency end the performance is poor because of shunt capacitances incurred within cross-point arrays at the exchanges. Leased analogue lines are hard wired through exchanges, avoiding any switching frames. There is a consequent increase in bandwidth, allowing higher signalling rates. Another advantage with the avoidance of switching frames is the reduction of crosstalk which also occurs. A higher bandwidth combined with reduced interference and crosstalk allows more sophisticated modulation schemes to be employed.

2. Kilostream requires a digital leased line to be connected between the user's premises and the digital exchange. The telecommunications carrier reserves pathways through their network and an identical Kilostream link to that described is needed for the remote end. Once these are in place a virtual permanent circuit is available to the user. This is similar to a leased line in the PSTN, in that the service offered is not flexible, i.e. communication is between two fixed sites.

   The digital line at each end must be terminated with a network terminating unit (NTU). The NTU interfaces the digital line from the exchange to the X.21 or X.21bis interface expected in the terminal or host computer. The packet switched stream requires everything that is necessary for a Kilostream link. In addition a packet assembler disassembler (PAD) is needed on the user's side of the NTU. The carrier treats the data as packet entities, and individual packets transmitted to the exchange may have different destinations. The packet switched stream (PSS) is thus more flexible and allows data to be delivered to any destination having an X.25 address. Indeed, X.25 allows closed user groups so that a virtual network could be created over a wide area.

3. Varying a sinusoid in some way is the basis of modem technology. A wave can be modulated in terms of its amplitude. It may also be varied in frequency or it may be subjected to instantaneous phase advances. Sinusoids may carry logical 0s or 1s by representing them as two amplitudes, two frequencies or two relative phases. Some systems use combinatorial modulation schemes, changing two parameters in order to signal more than a single states.

4. Scramblers and descramblers are often thought to be used to provide security. This form of scrambling could be achieved by 'jumbling' the frequency spectrum. While this might be a legitimate definition in telecommunications generally there is a specific meaning and function of scrambling and descrambling in modems which operate with phase shifts. The modem must maintain a reference oscillator in order to recognise phase shifts. The possibility of a number of sequential 0000 or 1111 quadbit signals would just provide a single level output from the demodulator. The reference oscillator at the receiver would not be given any synchronising information from such a level. The concept of scrambling is employed in order to ensure that although such runs may exist as data, they cannot exist in the signalling.

5. In the integrated services digital network (ISDN) the customer is provided with a digital local line from the digital exchange. The exchange line must be terminated at the customers premises by a unit called an NT1. The NT1 (network terminator type 1) supplies the interface to the customer from the line which may carry 2B + D data at 192 kb/s. The NT1 is a single output device. If a number of customers' devices are to be connected then another unit can be added. This second unit is called an NT2 and it provides up to eight 'S' interface lines. NT2 uses collision resolution techniques to manage the inputs. S interfaces connect directly to terminal equipments that are fully compatible with ISDN. Such devices are called TE1s. Of course there may be need to connect other than true ISDN terminal equipments and the R interface is for such cases, when used with a terminal adapter. The NT2 is able to match the nature of the terminal and thus may provide, as one of its manifestations, LAN controller functions with the appropriate protocol structures. The 'U' interface is specific to the carrier's network and is thus irrelevant to the customer. This discussion has been based upon a standard rate connection and described the B channel, digital 64 Kb/s and the D channel, digital 16 Kb/s. Additionally, ISDN services would seek to provide A channels for analogue voice telephone connection and E channels designed for 64 Kb/s packet switching. There are also a number of H channels designed for multiple links but these options need the primary rate interface, e.g. H12 which gives 24 B channels yielding a total data rate of 1.536 Mb/s.

Chapter 7 Going mobile

1. The main factor that constrains how radio and satellite systems operate is the distance between nodes. Satellite communication systems operate over greater distances than any terrestrial link and so they might be expected to exaggerate the problem making it easier to grasp. With a geosynchronous transponding satellite, the slant range from a ground station will be in excess of the satellite altitude and thus a message

travelling up to the satellite and subsequently travelling down to a remote ground node will take more than 240 mS. Consider the effect this would have on a polling access for a population of say, 100 ground stations. Each poll, if it worked on the HDLC basis would require a poll message and a reply. It would take about 50 seconds for a single poll cycle. This is unacceptably long, but this assumes a ground station in charge of the polling, e.g. a ground master. If the master resides in the satellite there is an improvement but at best the fastest 100 poll cycle will be 25 seconds.

Token passing requires that ground stations should pass the token from one to the next using the satellite as an intermediary. As a result the token circulation time is again 25 seconds. One more suitable option is time division multiplexing. Each ground station would be allocated a time slot within which to transmit. The disadvantage is the inefficient use of the spectrum when ground stations are idle and the difficulty in accommodating network growth. If traffic offered by ground stations is relatively sparse then contention broadcast would also be suitable. The only problem is that the window of possible contention is very long and there might be a tendency to instability (lock up) during periods when some station offer a lot of traffic. A compromise is to operate as a contention bus equivalent, taking advantage of the perfect feedback inherent in radio systems. In the event of lock up the system could detect the lack of throughput or the continuous data collisions and switch to a time division multiplexing mode which will eventually clear the log jam. The proposition has a strong element of truth in it.

2. The differences are considerable and stem from a fundamental difference in channel space allocation. The analogue system (TACS) uses frequency division multiplex (FDM) while the digital system (GSM) uses time division multiplex (TDM). GSM achieves the same traffic capacity with an eighth of the number of carriers. Rather than use the large number of analogue filters present in TACS hardware, GSM uses fast time domain switching to separate stations. Whereas TACS transmits an analogue speech channel, GSM generates a digital sampled data system speech channel.

3. The problem with cellular radio is that it has been developed by telecommunications authorities on the basis that the majority of links will be carrying speech between people. This assumption has proved true but the situation might be self fulfilling, since most of the marketing is aimed at speech applications. Cellular radio is optimised for speech and so the simplest method of reliably transmitting data is to use the methods devised for the transmission of data over telephone wire circuits. Any modem technology should be effective.

4. The main justification for radio networks in usage terms is the need for mobility. Obviously there is no option when it is necessary to communicate with moving vehicles. On the geographical side, radio can be used as a cheaper alternative when population density is low and the expense of

installing optical fibre or copper wire is not justified because few other subscribers are present. In this case even fixed installations are well served by radio. Some remote regions also have geographical features that would cause problems for VHF or UHF terrestrial systems. Typical situations might include locations in deep valleys between mountain ranges. Quite modest hills can act as barriers for radio propagation in the frequency ranges permitted and for these situations a satellite communications system is likely to be effective because of the usually high angle of the satellite from the ground. VHF and UHF low power transmissions are limited to approximately line of sight working and a few problems could arise from the earth's curvature. Ships or boats can be below the radio horizon at these frequencies when they are a few tens of miles out at sea. Satellite communication is a natural solution.

5. The flooding notion is based upon the idea that a mobile station might be anywhere and in order to guarantee that they reach it messages should be forwarded to every single relay node. The flooding routing strategy will do this. In reality, many radio systems are much better informed. Cellular radio incorporates a 'refer back' which checks out a 'foreign' node and authenticates that the node is valid and the services that it may receive. The process is useful from the system administration point of view but it also allows the home base station associated with the node to 'know' the location of the node and thus route messages to it if necessary. Flooding would not be justified with a cellular system, but it might be in a private radio relay system which did not need sophisticated tracking and accounting requirement.

Chapter 8 The company network

1. The suggested network is national in scope and although there are four sites mentioned together, with an indication that each site creates significant amounts of data, it is not clear whether each site generates more or less traffic than any other. There is no time profile to help identify whether 'regular' means every 15 minutes or for 20 minutes every day at two in the morning. In cases of this kind it is necessary to make assumptions in order to proceed. It will be assumed that regular implies almost continuous traffic. With the public packet switched option, charges include a per packet element. Alternatively, a digital leased line is charged on a per link annual basis. No private company can expect to own links of the length required here and consequently a public carrier's telecommunications network will have to be used and for constant high density traffic a Kilostream or Megastream option would be chosen. This is also appropriate since it is an equivalent leased line option connecting fixed sites. The question now arises as to how data is handled over the leased line system. The carrier is not concerned with details of the protocol and thus the company network management may decide upon the details. To enable the links to carry a number of independent virtual channels a good choice

would be private X25. Multiway PADS could be installed at each site, and dependent on the precise topology for the network, a packet switch with sufficient input and output buffers could be installed. The company's network management would be responsible for the configuration of the switch and the PADS and the company would have to purchase these items. Only the digital data highways between the sites would be provided by the carrier.

2. Company networks for intra site operation must be chosen with due regard for the context of their use and the appropriateness of the technology choices. General machine control in a motor vehicle factory implies the use of the network to carry programs and data for numerically controlled milling and turning centres and for robots and manufacturing cells. One key point about this kind of installation is the absolute need for certain messages to be delivered in a guaranteed maximum time. To appreciate this, consider an emergency stop signal to a powerful industrial press or milling machine. If the message does not get through very quickly then a lot of damage will be done. This kind of constraint leads to the prescription that only a deterministic network is suitable. One well known deterministic system that has been devised for just this application is the Manufacturing Automation Protocol (MAP). It consists of a token bus operating on a basic broadband medium. Only two channels are implemented, a seperate transmit and receive carrier for the nodes, translated by a head end unit. The carriers operate around 600MHz to avoid the industrial noise spectrum. A big plus for this system is that it has been continuously developed during its lifetime to date and it does have a large number of installed sites which leads to the economies of large scale node production together with an active user group community. MAP is based fully on the open systems ISO 7 layer model. The recommended system would work in a drawing office environment but the constraints are far less demanding. Office environments are far more friendly than manufacturing production line conditions. MAP would be unnecessary in the drawing office and thus unnecessarily expensive. There is no requirement for worst case delivery times when computer graphics is the main traffic area. Consequently, a CSMA/CD probabalistic system would be appropriate. Naturally, if the drawing office wished to download design data directly to manufacturing machines then there would have to be compatibility between the two networks. The compatibility does not have to be literal but can be provided by means of a bridge or gateway.

3. Building Society offices vary enormously in size and location. Some are as busy as a city bank, others are located in very small towns or even hosted within an estate agent's office. The traffic generated will differ considerably for each case. Many rural places in europe do not yet offer digital line services. Such rural places tend to contain the offices generating least traffic and it may be that the service provided by dial up modem over the public switched telephone network would be adequate. The busiest city centre sites might well form a backbone of 64kb/s leased digital lines using

private packet switching. For intermediate offices, where digital lines are available, and where the traffic generated is not of the continuous type then a link using X25 over the public data network is satisfactory. Generally, calculations based on accurate traffic projections and cross referenced to the services and charges offered are the starting point.

4. PACX stands for private automatic contention exchange. Such a device is concerned with circuit switching. It is the equivalent of a telephone exchange for data circuits. A network could be constructed by means of links via leased lines, either analogue or digital, interconnecting switching centre sites. The 'switch' at such sites is an exchange based on crosspoints and time division mutiplexing. The number of designers specifying switches of this kind is diminishing fast, the reason being the popularity of packet switching and the relative demise of circuit switching. The PACX usually uses groups of switches in order to reduce the number of physical switches required. Often, a three stage process is used with a concentrator, distributor and expandor. Assumptions are made about concurrent usage and as a result there are fewer pathways through the switch than there are circuits connected to it. Clearly, the assumption is that not all users will be active at any one time. This is reasonable, but when the exchange is fully in use, the next circuit demanding attention will have to fight with others for the privilege of connection. This is the 'contention' referred to in the PACX name.

5. Companies are always advancing with the state of computing technology and notable examples would be the use of multimedia or video conferencing. Such development rely on the ability of networks to carry a mixed economy of signals. At any time signals might be graphics images, speech, textual data, or remote procedure calls. ISDN was designed from the outset to accommodate the variety of signals to be expected. Although ISDN was devised and worked up by telecommunications carriers there are many interfaces vendors. In the same way that private X25 can be operated by a company, there is no reason why a company could not use ISDN technology over its private network. The only necessary prerequisite is that the links and data highways used must have sufficient bandwidth.

Chapter 9 Communications data processing

1. The answer depends upon whether the traditional discipline of pattern recognition is being narrowly adhered to or whether a more liberal interpretation is possible. In the latter case there is an element of pattern recognition in the Ziv-Lempel algorithms. In LZ77, digraphs and trigraphs in a look ahead buffer are searched for in a past data window. On recognition, pointers are used in lieu of the letter group in question. This example operates on textual, perhaps ASCII, characters. In a more traditional sense, objects in images or characters in a document might be segmented from the image and compared with library entities by a pattern

classifier. The result would be the ability to signal a library identification number instead of the object. Naturally, both ends of the links must maintain identical libraries and will have to accommodate methods of adding to the library and updating. The first case described is redundancy reducing and noiseless. In the second example the classifier may have a statistical nature and would be unlikely to be able to fully generalise the slight variations that might exist, say between two characters intended to be the same but one of which is slightly corrupted. The level of tolerance exhibited by the classifier will affect the character of the compression. A neural net does generalise to an extent and so in the special case where it intended that small differences should exist between two essentially similar objects then a small dissimilarity might be ignored. This would be an entropy reducing situation.

2. Appropriate methods are application specific. The nature of maps suggests that they might be stored as 'quadtrees'. A quadtree is a hierarchical representation of non redundant pixels which is based on an array. A portion of the map is first divided into quadrants which are then subdivided to the depth required. The quadrant numbers at the top level may refer to a block. At the next level down a quadrant refers to a block of a quarter of the size of the original and so on. With a depth of four the top left hand block might be referenced as 1000. The top left hand sub quadrant would be 1100, the bottom right hand sub quadrant of 1100 would be 1130 and so on. The representation allows the storage and examination of data on a 'gross information first' basis. A map could be examined at low resolution by reconstituting 1XXX, 2XXX etc. An increase in resolution is possible by considering 11XX or 113X as required. Unfortunately, althought the representation is very useful for searching maps, it does not provide significant compression. Consider a list of non redundant pixels referenced:

   1212,1213,1223,1232,1233,1234,1240,1320,1330,1340 and 1400.

   This could be expressed as a starting value followed by differences. The same list would be: 1212,1,10,9,1,1,6,80,10,10,60. Additionally, special codes for repeat sequences, and blank quadrants could utilise digits above 4. Experiments with this notion have yielded compressions between 3 and 32:1.

3. The purpose of the transform in transform coding is purely to decorrelate a series of samples. Many data sources generate consecutive pixel entities that are highly correlated. For signals that are a function of time, transformation into the frequency domain is effective here. The transformation does not generate compression. It merely makes the data more amenable to compression by decorrelating the samples. Compression in these systems is provided by the quantisation process. Quantisation is the allocation of a given number of bits to data which in its raw state comprises a much larger number of bits. This is the entropy reducing process. It is usually achieved by setting a threshold amplitude and zeroing values

below or selecting a range of frequencies considered more significant than others. The samples defined as insignificant are similarly zeroed. Transformation into the frequency domain is particularly effective because it tends to concentrate high amplitude components at the beginning of the vector (1-D) or the top left of the array (2-D). A large time domain vector of sample values may be reduced to a few significant values and a run length of zero values. Compression is thus implemented. The joint picture experts group (JPEG) lossy algorithm employs the discrete cosine transform (DCT) rather than a simple fast Fourier option. In the 2-D model it weights the array to assist in simplifying quantisation. It also reads off the array in a zig-zag starting in the top left. This further concentrates significant samples at the beginning of the list. Run length coding compresses the zeroed values. If moving pictures are being compressed then only changes between sequential images are considered.

4. Ziv and Lempel devised two compression algorithms based on string matching techniques which take advantage of the redundancy inherent in repeated or repetitive strings of textual characters. LZ77 uses a sliding history buffer. This window allows new data to be entered at one end and the oldest data is ejected from the other. If the buffer were 2K characters, then the last 2K would always be 'in stock'. A look ahead buffer is examined and character string matches of various lengths are searched for within the window. Control of the search is by pointers and when a string match is found the pointer reference and a string size are substituted for the repeated string. The factors affecting the performance of LZ77 are: the size of the window; and linked to this the pointer size . If the window is too small then string matches will be less likely. If it is too large the pointer size becomes sufficiently great to affect the compression performance.

   LZ78 is based on a tree structure. The tree is really a form of 'dictionary' in which new words are attached to a root node sharing a common starting sequence, e.g. 'parrot' and 'parallel' would share the same root node 'p', and subsequently branching only after 'ar'. The end of the branch in each case is given a reference number and since the decoder must have the same 'tree' then only the reference numbers are signalled and the receiving end recovers the decompressed word by starting at the reference number leaf node and tracking back to the root. Factors affecting the performance of this algorithm are: the size of memory available for tree constructions; the efficiency of the searching algorithm used; and the system response once the dictionary has become full. In the event of this, it is clearly necessary to monitor the compression ratio being acheived and should the data source statistics change then a decision must be taken about whether to start again with a new dictionary.

5. The data encryption standard (DES) need not be implemented as a block cipher. It is possible to operate the algorithm in a 'stream encryption mode'. The exact technique is outside the scope of this book but a general appreciation can be given. There are parallels between the method and the

idea of CRC. The sending end creates the ciphertext by XORing the plaintext with the output of the encryption system. The encryption system includes a shift register through which the ciphertext is processed as it is generated. This produces a historical feedback which alters the subsequent encoding. In this way, the repetition of a fragment of plaintext never produces the same ciphertext. At the receiving end the same encoding algorithm is applied, this time with the incoming ciphertext XORed with a processed version of itself. It is intriguing to consider the result. The plaintext character being encrypted is XORed with the same processed output that is XORed with the arriving ciphertext at the receiver. The output is the plaintext. Assuming the same initial conditions at both ends of the link then the system encrypts, sends and receives each character as it is generated. The DES process for this action may be created purely in hardware and implemented as a VLSI chip. Such chips may be used in real time data communications links.

Chapter 10 Network management

1. The object of network management is to provide a transparent and effective service to users. A common test is to ask whether the service that a user receives is as good as he or she would have had if a stand alone system carrying all the necessary software were at his or her disposal. System security is an important management issue. Security may be tackled at many levels, from ensuring that only authorised users are able to access the system to ensuring file access is controlled. It may even be necessary to take measures to minimise eavesdropping by outsiders. The tendency has been for network managers to be quick to install controls but slow to create smooth, quick and effective systems for operating changes to the security aspects. Users become rapidly disenchanted if it takes a week to gain access rights to specific data. The 'stand alone' equivalent test will be failed. Another significant issue is fault management and recovery. The manager needs help here, although in many organisations the function relies on one or two knowledgeable individuals. A sensible approach is to employ monitoring middleware which is able rapidly to provide babysitting and post/pre fault tracking of events. Workload management is key to the new world of corporate network computing. If corporate production jobs such as payroll must be processed alongside the casual use of network bandwidth then it is essential to have detailed monitoring and daily profiling of the network usage. Management tools offering 'what if' schedule trialling are invaluable.

2. The fundamental network monitoring tool is the protocol analyser. The data recording function enables the detailed gathering of information on traffic offered, delays, sources of network hogging, real time profiling by skyline diagram, number of errors, number of data collisions in a contention system, incidence of low level errors such as misaligned frames, runts and jabbering. Without information of this kind the network manager is

helpless. Other useful instrumentation may be provided by functions of the network operating system. Typical examples might be user accounting data or resource utilisation statistics, e.g. average queue sizes and printer spooling delays.

3. One of the main issues in security is the system's ability for global access and asset control. Files must be capable of being made 'read only' to some users. Application programs should be available only to those with permission to run them. This is necessary to avoid accidental or even deliberate data corruption or copyright breach. If a user were able to download the executable code for an application there would be clear copyright problems. Access to data and programs may be assigned by means of the validity of the user's name and may be enhanced by the use of passwords. Other useful features include: enforcement actions coupled with monitoring so that say three tries to beat the password causes lockout; monitoring of real time warning messages relayed to the system console; and calendar/time based access control so that data cannot be accessed outside normal office hours.

4. Printer spooling is necessary because a printer is a shared resource. Random access cannot be allowed because of the danger of intermingled print output. Spooling is a queue process, based on a first in, first out system. Documents for printing are this lined up one after the other in a print job schedule. In a large multiuser machine, memory can be allocated to the print queue and queue control programs can run as one of the round robin tasks. With this a single centralised queue can be resident in the main memory. A networked system is different, in the sense that many users may call for print services even though they are geographically disparate and individually have no knowledge of other callers. This can be handled by the network operating system blocking other contenders or by running a print queue in the print server machine. The last option is typical.

5. Computer virus infections are anathema to network managers. A discipline must prevail within the company which ensures that no unauthorised software is loaded onto users' local machines. The presence of the network and the cleverness of some virusses would allow cross infection and a rapid degeneration of the system. However strong the discipline or the sanctions there is still a risk and many managers have acquired comprehensive virus search and identify packages. When a user logs on, the search package checks out all the files in local backing store of the user and looks for more subtle viruses and trojans. This strategy has been very effective but relies heavily on first class updating and maintenance of the virus/trojan search package. If the package finds something it is unhappy with it causes the network operating system to block the user from the network and require him or her to call the network management to have their system purged.

Chapter 11 Distributed systems

1. A network operating system expects that the computers in the network will have mixed origins. They might all be basically of one type but they will have been bought at different times and will have differences in their BIOS (basic input output system), and might be running different versions of the operating system. There could be many differences. A network operating system expects each machine to run its own operating system which looks after the local allocation of resources, file management and so on. It adds processes which act as interfaces to the network. These interfacing processes are called 'agent processes'.

   A distributed operating system expects all machine to be running the same operating system kernel. Generally there is a 'same hardware specification' requirement. In a distributed system the network should be invisible. It should be as if only the local processor exists and any accesses to files elsewhere or the allocation of remote resources are completely automatic.

2. A parallel language expects to be able to distribute tasks to central processors other than the local one. In order for another remote processor to execute a partial program it is necessary for the program object code to be downloaded and there should also be the ability to pass parameters and trigger program execution. Clearly, interprocess communication systems are a necessary underlying capability.

3. If a remote processor crashes while executing a distributed program there would be a danger of an indefinite hold up in the overall job. Fail-safe or fail-soft mechanisms seek to use the systematic application of message logging and progress logging. Additionally, back up copies of messages should be kept so that in the event of a problem tasks could be restarted or even redistributed to alternative processors.

4. Many organisations have been developed to satisfy the need for parallel computing. There are multiprocessor systems, which tend to be tightly coupled arrangements where a relatively small number of processors share a common memory. There are massively parallel systems in which a large number of processing elements are each provided with dedicated memory and in which separable tasks are distributed for parallel execution. The processors are all co-located. Another category is the multicomputer system where a large number of autonomous machines are co-located and connected via a high speed communication link. Apart from these there is one further option, which is where autonomous computers are connected by means of a local area network. The processors need not be in the same location, in the sense that individuals might be geographically separated by up to several kilometres. In this last case a possible paradigm of operation could be the distributed operating system. Each processor would run an identical copy of an operating system kernel for which an important element is a message passing or remote procedure calling

capability. The extra complexity is justified with any application that would benefit from the parallel process execution that is possible. Matrix or vector problem solving is such an application. Another option is the distributed database system.

5. In a distributed relational database the data is stored in tables which are known as relations. The columns in these tables are called attributes and the rows are called tuples. In a distributed database one of the basic decisions is to determine how to distribute the relations. Many factors need to be considered, including:
   - the number of relations in total; the pattern of activity expected;
   - the number of processors expected to access the data;
   - the need for copies and the storage facilities available;
   - the communications traffic expected to be offered within the pattern of operation.

   It can be appropriate to distribute the whole relation by replicating it at several sites, or sometimes it is more efficient to break down the relation and keep subsets of the relation in different places. A subset of attributes (columns), when distributed, is an example of vertical partitioning of the relation. Similarly, if a relation is broken down in such a way that different rows are stored in different places then horizontal partitioning has been implemented.

Index